ELEMENTAL ANALYSIS IN GEOCHEMISTRY
A. MAJOR ELEMENTS

SERIES

Methods in Geochemistry and Geophysics

Methods in Geochemistry and Geophysics

8

ELEMENTAL ANALYSIS IN GEOCHEMISTRY

A. MAJOR ELEMENTS

BY

ALEXIS VOLBORTH

Killam Professor of Geology and Mineralogy,
Dalhousie University,
Halifax, N.S., Canada
Formerly:
Mineralogist, State of Nevada,
Nevada Mining Analytical Laboratory
and Professor
University of Nevada, Reno, Nev., U.S.A.

ELSEVIER PUBLISHING COMPANY
AMSTERDAM / LONDON / NEW YORK
1969

ELSEVIER PUBLISHING COMPANY
335 JAN VAN GALENSTRAAT
P. O. BOX 211, AMSTERDAM, THE NETHERLANDS

ELSEVIER PUBLISHING COMPANY LTD.
BARKING, ESSEX, ENGLAND

AMERICAN ELSEVIER PUBLISHING COMPANY, INC.
52 VANDERBILT AVENUE
NEW YORK, NEW YORK 10017

STANDARD BOOK NUMBER 444-40711-1
LIBRARY OF CONGRESS CARD NUMBER 68-29647

WITH 76 ILLUSTRATIONS AND 15 TABLES

PRINTED IN THE NETHERLANDS

Preface

After the pioneering compilation work of Washington and Clarke on major elements, the geochemical research of the past has been conducted predominantly on trace elements, so that for some scientists "geochemistry" has become synonymous with trace element study. Today, the development of fast instrumental techniques has again focused our attention on the importance of major constituents. These developments create a need for the compilation of basic classical and modern instrumental "macro" methods with some emphasis on trace element determination, hence the title of this book which attempts to emphasize its dual nature. This task seems to be an impossible one considering the vast realm of analytical chemistry today, unless strict practical restrictions are applied. With this difficulty in mind, only a minimum of methods and theoretical background has been presented. The author believes that most essential determinations on a geochemical or complex inorganic sample can be performed with the aid of this text. By combining classical and modern instrumental methods in one book, strong emphasis is put on the importance of the analyst's ability to grasp the general structure and relation of some of the most frequently used analytical techniques.

The author is painfully aware of running the risk of oversimplification and showing a great degree of temerity in even attempting to perform this task. It must also be emphasized that the omission or selection of methods given in this book does not constitute criticism nor indicate superiority. The main criteria for selection have been the author's familiarity with the method and also the simplicity of actual steps required. It is not implied that the methods selected are the best or most straightforward, but it is implied that the methods given represent a collection of some of the best and easiest to understand and that they have been selected and presented with the widest possible application range in mind.

It is hoped that this work will permit the first step in the right direction for a beginner, that it will enable a geochemist or a chemist with a basically instrumental training to perform, when necessary, "wet-chemical" work, as well as induce the classical bench chemist to use fast instrumental tests to facilitate his task. An absolute minimum of information necessary to understand and actually do the work is compiled, and the analyst will soon need additional references which are given in the Bibliography. Direct literature quotations have purposely been omitted from the text for the sake of brevity and to enable the beginner to start and actually perform the work before asking questions. The author feels that a simple method well learned and understood forms the basis for further refinements and builds an *analyst* in the true sense of the word.

It is not implied that a complete analysis of complex mixtures can always be performed using this text, but it is implied that a variety of common silicates, carbonates, phosphates, sulfates, oxides, sulfides, and some metals, can thus be analyzed or at least a start in the right direction made before seeking additional specific information. In many cases, a combination or combinations of methods given should enable the beginner to make a rather complete analysis of a complex mixture.

It often takes more time and effort to abbreviate and give just the essentials than to describe a procedure in detail. In abstracting and abbreviating there is also the danger that the subject is presented in too terse a form. The author has attempted to insure sufficient brevity while saying enough by consulting a group of experienced analysts who also have read the manuscript, especially in parts relating closely to their field of specialization. In cases where my own experience was insufficient to produce a brief yet practically useful description, I was able to visit scientists in their laboratories to observe and participate in new analytical techniques in order to gain first-hand experience. This part of the work was performed in

1965–1966 as a John Simon Guggenheim Foundation fellow while on sabbatical leave from the University of Nevada.

For the past three years, during the compilation and reduction of data for this book, Dr. H. A. Vincent, Chemist, University of Nevada, and Mr. Matti Tavela, Geochemist, California Division of Mines, have generously given time and effort by criticizing, discussing, and reading in detail most of the chapters of this work.

This book is basically a manual of methods, and contains some instrumental methods partially developed by the author at the University of Nevada, Nevada Mining Analytical Laboratory, during the past ten years.

In addition to the University of Nevada where the major part of the analytical work has been performed since 1956, part of the experience in modern methods has been gained while visiting other institutions. Professors A. K. Baird, D. B. McIntyre and B. L. Henke from the Departments of Geology and Physics, Seaver Laboratory, Pomona College, have introduced this author to the ultra-soft X-ray techniques with emphasis on statistical interpretation of results. Dr. H. E. Banta has taught me the practical aspects of fast-neutron activation at the Oak Ridge Institute of Nuclear Studies. Professor Hidekata Shibata, Tokio University of Education, has personally demonstrated his systematic ion exchange analytical method for silicate rocks. Dr. R. W. Nesbitt, Senior Lecturer, Geology Department, University of Adelaide, Australia, and Mr. P. A. Weyler, University of Nevada, have introduced me to the routine of atomic absorption. Dr. K. Norrish, Commonwealth Scientific and Industrial Research Organization, Division of Soils, Adelaide, has kindly provided detailed information on the X-ray emission analytical methods developed by him, and has permitted the use of previously unpublished material for this text.

Parts of this book have been written at the University of Rome, the University of Adelaide, and the Tokio University of Education. Professor Mario Fornaseri of the Istituto di Geochimica, Rome, has greatly contributed by inviting the author to use the unique and modern geochemical library at his institute, and by giving much of his time in the discussion of the relative usefullness of certain analytical methods.

Chemists and geochemists at the U.S. Geological Survey, as well as the scientists mentioned above, have read the manuscript or portions thereof, and have made numerous valuable suggestions. Especially, the author is thankful to Messrs. Michael Fleischer, F. J. Flanagan, F. S. Grimaldi, H. J. Rose, Irving May and L. C. Peck for criticism, advise, and help.

Much of the analytical work which formed the basis for this book has been performed by the author while working under the auspices and with financial support of the National Science Foundation and the U. S. Atomic Energy Commission (NSF Grants GP-864, GP-1987, GP-2222, and GP-5371; and the Project Shoal, Vela Uniform, A.E.C.). These United States Agencies have contributed greatly in the establishment of the fast-neutron activation and the X-ray facilities of the Nevada Mining Analytical Laboratory as a modern institute for accurate major and trace element analysis of natural and industrial substances. This laboratory is presented in this book as an example of a combination for classical and instru-mental nondestructive analysis.

The expert drafting of the figures was done in part by Mr. R. R. Paul and John L. Murchie. The typing of the manuscript has been performed by my wife, Nadja Volborth, and by Gail McGillis, Research Assistant. Gail McGillis has also initially edited and proof-read the manuscript.

While taking full responsibility for the contents of this book, I am greatly indebted to all the persons mentioned.

ALEXIS VOLBORTH

Contents

CHAPTER 1

Introduction

In the age of revolutionary development of instrumental techniques of elemental analysis, the bonds between this branch of inorganic analytical chemistry and the classical gravimetric and volumetric methods tend to weaken. This affects the thinking of the young chemist whose concept of basic knowledge, which is necessary to perform and evaluate an analysis, becomes biased, which in turn results in impaired ability to perform an *analysis* as contrasted with plain *determination*.

Working with instruments and producing a multitude of data can create a wrong sense of accomplishment and confidence. The beginner too easily becomes a button pusher, a "determinator" rather than first an *analyst*. This leads all too often to unwarranted aloofness and dangerous separation of the instrumental methods of analysis from the chemical separations which still form the backbone of analysis of complex mixtures. Irrespective of the method used, most instrumental and even the so-called nondestructive techniques such as neutron activation and X-ray emission, for example, require some form of classical chemical attack of the sample and often extensive separation pre-treatments before the element or elements can be determined.

Paradoxically, the rapid development of instrumental techniques seems to have made the accurate classical approach less fashionable, whereas the evaluation of data produced by these new methods still depends to a great degree on the ability of the chemist to perform preliminary group separations based on systematic classical analysis. Yet, the need for complex analysis has increased in proportion with technological advances and the use of new ores and materials. We have only to mention the problems encountered in the analysis of reactor materials and spent radioactive fuel elements, to realize how immensely complicated the task of an inorganic analyst has become today.

Geochemical research, closely connected with the search for new sources of raw materials, requires more and more complete analytical data of rocks, ores, minerals, and industrial substances. It thus becomes necessary to check the accuracy of one determination by performing a series of determinations or even perhaps a complete analysis, to correct for interferences or to insure that no constituent has been missed.

The intention is to acquaint the beginning analyst or geochemist with the minimum of analytical methods required to satisfactorily perform a complex silicate or similar analysis. The individual procedures or steps given will also permit a single determination of an element in a complex mixture if separation methods described are followed. Elements for which methods are given are either major elements of the earth's crust or trace elements which have basic industrial importance and which occur in major concentrations in their ores or materials. Trace elements in the geochemical sense, which are also industrially uncommon, will be treated in another volume. However, the instrumental methods as described are generally applicable to trace element analysis as well, if proper precautions are taken.

Due to the dual nature of this text there was some question concerning the inclusion of certain chapters into this (first) volume. It may therefore be felt by some readers that a chapter on standards, a chapter on optical emission spectrography, and a more exhaustive treatment of selection of representative sample- and grain-sizes are missing. Also, some excellent colorimetric (Cu) and distillation methods (F) suitable for trace element as well as major element work are not included. This is because these subjects were considered more related to trace element analysis, for which a second volume is in preparation.

This book is designed to enable the beginner to follow logically through sample preparation, statistical evaluation and the consecutive qualitative and quantitative schemes till the "total composition" of an unknown sample is satisfactorily known. Two systems are built in—one *destructive classical* and one *nondestructive instrumental*. The methods are selected so as to overlap as fully as feasible in order to enable the chemist to produce at least a "dual" set of independently retrieved data for each element in order to check the accuracy of results and detect possible methodical errors. To achieve this, the major elements are treated consecutively in the usual sequence of estimation according to the methods used. Special procedures and steps are cross-indexed to avoid duplication.

As a practical example of how to use this book, we consider the following case:

Constituents of a rock powder: O, Si, Al, Fe^{2+}, Fe^{3+}, Ca, Mg, K, Na, Ti, Mn, F and P are sought. At least two independently obtained results are required for each of the elements listed. The rock is to become a standard.

(*1*) Starting with Si, one determines, in the sequence listed, all the elements, except oxygen, by the classical gravimetric and volumetric procedures, after fluxing with Na_2CO_3, and $CaCO_3$ (pp. 65, 91).

(*2*) One determines Al, total iron as Fe^{3+}, Ca, Mg, Mn, Ti, K, Na and P according to Shibata's method (p. 58). This can be done first in order to get a general composition of the sample.

(*3*) One determines total Fe and all the elements listed (excluding F) by one of the X-ray fluorescent techniques (O by ultra-soft X-rays, if available).

(*4*) One determines O and Si (also F) by the fast-neutron activation techniques, which are very accurate for these elements.

(*5*) One determines Mg (and some of the other elements listed if desired) by atomic absorption.

This combination of methods gives a minimum of two results (except for Fe^{2+}) for each of the elements, permitting a selection of best data and detection of methodical errors. A careful analysis of an unknown sample by this combination of methods can be expected to yield significant results establishing such a sample as a useful secondary standard.

Going a step further an analyst will soon be able to establish which of the methods gives most reliable results with least effort and time consumed. Applying proper corrections to instrumental techniques, an analyst will then be able to shift the emphasis from time consuming classical to rapid instrumental techniques whenever possible without sacrificing accuracy or reliability of results. Assuming that the necessary equipment is available, one ends up ideally with a minimum of gravimetric and other chemical separation techniques and an emphasis on computerized rapid non-destructive methods retaining, nevertheless, the continuous capacity for double checks by the classical accurate methods, whenever doubts may arise.

CHAPTER 2

Preparation and Decomposition of Samples

PREPARATION OF SAMPLES

This book attempts to cover the elementary problems encountered in analysis of silicate, carbonate, sulfate, and phosphate rocks, sediments, minerals, ores, portland cement, glasses, glazes, slags, ashes and some common ferrous and nonferrous alloys. This list of materials presents a very wide variety of samples which makes a detailed description of standardized or common sampling procedures impossible. Some of the general principles and considerations given below will, however, orient the beginner and may help him to a meaningful start.

Common types of samples and methods of sampling

The main purpose of all sampling is to obtain a reasonable quantity of the substance which would contain all the constituents present in the bulk mass and in the same relative amounts. When this is achieved the product is called a *representative sample.* It is obvious that homogeneous substances will require smaller samples than heterogeneous mixtures. For example, in glasses, mineral waters, or liquids, and in single homogeneous phases of minerals, the size of a representative sample is controlled by the most convenient quantity of the material taken from the analytical standpoint, whereas in some pegmatitic rocks, layered sediments, or migmatites, tons of crushed and homogenized material are necessary to obtain the bulk composition representative of the whole rock. Unless the sample is properly taken, little can be gained by spending time and effort to determine its composition. This is especially true in relation to geochemical studies of rocks. High demands of the present-day technology make accuracy also crucial in industrial analysis. Only when a preliminary and approximate composition is desired, as in the case of original "grab" samples of ores, is it feasible to analyze quantitatively such unrepresentative samples at all. Many geologists consider any and every sample representative and argue that otherwise one would need to analyze the whole rock body.

In a strict sense, however, the size of a representative homogeneous sample can only be determined by statistical analysis of some analytical data. A main constituent, such as silicon, can be chosen for this purpose. This method is feasible when the analysis is performed by the fast nondestructive X-ray fluorescent techniques as outlined elsewhere in this book. When it can be shown by repeated determinations that increasing the size of the original sample does not alter the result more than within the usual standard deviation of the analytical method, one may assume that the sample is representative. This approach to the selection of the size of a representative sample from a single location is valid only with rocks and solids of relatively uniform character and occurring in great masses, such as igneous plutons or lava flows for example.

In very heterogeneous materials such as migmatites, pegmatites, hydrothermally altered rocks pierced with vein material of very extreme composition, conglomerates, agglomerates, volcanic tuffs containing mafic as well as siliceous fragments, greywackes, gravels, and other heterogeneous sediments, one usually constructs a suitable grid system over the area to be studied. One then can take equal samples of relatively small size from randomly selected numbered intersections. These are combined into one representative sample, homogenized and analyzed. For example, four such composite samples consisting of eight random samples each may be selected. Naturally, each of these samples, if first analyzed, separately will indicate variability of the rock.

This is called *sampling by increment*. Each sample in this system represents the composition at a specific place, and a sufficient amount of samples combined is assumed to be representative of the whole rock. It can be shown that this approach gives a representative sample considerably faster than if one simply attempts to increase the sample, which would lead in many cases to insurmountable difficulties with enormous sample quantities without giving reliable results.

Inhomogeneity of rocks causes special problems in determining the size of a representative sample because this will depend at least on the element, the grain size of the mineral or minerals in which the element occurs, and the concentration level at which this element occurs in the rock. Whereas major elements may be sufficiently represented in a 2-pound sample, some trace or minor elements occuring concentrated in a few scattered mineral grains, may require a much bigger sample. As classical examples of such cases, we may mention zirconium in zircon of the granites, and native gold in disseminated ores.

Sampling crushed ores and slags from evenly spread piles is ideal but can be a very difficult problem due to the inaccessibility of the whole body. These materials are often piled up and are inhomogeneous, which further complicates the sampling. But in general, unless one can sample during loading or unloading, one tries to form a concept of the shape of the pile and takes samples also from within the body without regard to the type of material encountered or personal preference, and tries to get the fines in the true proportion to the general volume of the body to be sampled. One must consider also the effects of transportation on materials of different density, grain size, and shape.

Transportation may cause segregation of heavy as well as coarse and flaky particles and one should consider moisture absorbed or lost during transit when sampling fine powders, for otherwise serious errors could result. The size of the single sample can be reduced when powders are used.

Alloys can be sampled by drilling randomly and combining the chips. Some alloys and steels present major problems to a sampler. Surface oxidation, oils, paints and all accidentally altered, heated or etched regions are to be avoided.

Many commercial products such as cements, glasses, glazes, ashes, pure metals and some alloys are homogeneous enough so that taking of a representative sample should present no difficulties. But nevertheless, in sampling nothing should ever be taken for granted. It always helps to examine the material carefully, study it in thin-section or polished section under the microscope, and examine the grain size, signs of alteration, and learn about its origin and properties as much as possible.

It is very difficult to discard personal bias when sampling. Often one subconsciously reaches for the piece that stands out, has something unusual in its appearance or in general seems to correspond to the sampler's preconceived idea about the rock or the ore he is investigating. Therefore it is not wrong to let somebody disinterested do the actual sampling according to a selected random method, only making sure that meticulous care is used.

In selecting the areas to be sampled often the first bias is already introduced, for example by selecting the exposed fresh-looking contact areas and ignoring the difficultly accessible, somewhat more deeply weathered and probably more tectonically disturbed or hydrothermally altered areas or vice versa. Bias, or systematic error, can be of personal origin as well as of methodical origin. In both cases it causes individual results which all tend to be either higher or lower than the "true" value. Because of this, bias is difficult to detect statistically or to eliminate.

It is always advisable to take more sample than that judged necessary because carefully analyzed samples often become secondary standards and because the investigation is never really completed. Additional work on other elements or checks of previous analyses are frequently required.

Portable diamond drills providing cores of up to 2 m in length have been used with success in sampling plutonic rocks. Main advantages of drilling as against chipping with a hammer are: the uniform size and traversing nature of the core sample, easier access to unaltered rock, and the fact that rocks at preselected sampling points are more accessible even when the location is topographically unsuitable.

An old well-proven method applied in rock as well as in ore sampling is the *channel-sampling method*. One marks with chalk two parallel lines, up to 20 cm apart, traversing the area to be sampled in the wanted direction, and chips off samples along this channel to a certain depth by geologist's pick, or a portable drill at measured intervals irrespective of the nature of the encountered rock. One can also use a chisel and hammer.

Loose sedimentary rocks obviously present less problems in terms of effort. Some tricks may be employed. On vertically exposed surfaces, scratching an equally deep furrow with the sharp end of the geologist's pick and collecting the falling particles in a bag, gives a reasonably representative sample. On clays one cuts out a wedge or channel with a knife. Auger drills are used to obtain sub-surface samples of sands, clays, marls, and silt several meters deep. Sampling of *soils* is a specialized procedure outside of the scope of this book.

The collection of samples from the sea floor has recently been advanced with the development of special dredges, coring, and snapping devices. Finally the collecting of water samples from specific depths and waters containing different gases with concentration dependent on the temperature, pressure, and even the time of sampling, presents such specific difficulties, that it can only be mentioned here.

In general, the analyst has seldom the primary responsibility for the sampling methods used. It is usually the geologist, the geochemist, the mining engineer, or the metallurgical engineer who provides the original sample. In addition, national organizations responsible for standardization in different countries, prescribe specific sampling procedures for almost all commercial-type materials, so that one usually follows automatically in these cases the prescribed procedures and it would be futile to try to describe here any detailed methods for the multitude of materials covered.

Statistical evaluation principles of a representative sample are given in Chapter 3.

Sample size

Because of the considerations to be taken into account in sampling as outlined above, it is difficult to give an estimate of the size of a representative sample for rocks of different grain sizes. However, the following quantities will usually be sufficient:

0.5–1.0 kg: Fine grained equigranular rock, grain size less than 0.1 cm.

1.0–2.0 kg: Medium grained equigranular rock, grain size between 0.1 and 1.0 cm.

2.0–4.0 kg: Porphyritic rock, grain size between 1.0 and 3.0 cm.

4.0–6.0 kg: Coarse grained porphyritic rock, grain size between 3.0 and 5.0 cm.

> 6 kg: Coarse porphyritic rock, pegmatitic in part, grain size between 3.0 and 5.0 cm.

To determine the representative sample size one may proceed empirically as described above by performing statistical analysis of analytical data. Actually only rigorous geochemical investigations may require this approach, and generally one selects simply a larger sample to be "on the right side" or uses an empirical grain size chart similar to the one above.

Diminution of samples

Assuming that a sufficient representative sample is available, the next problem is to crush and grind it to a grain size suitable for further decomposition in order to make it amenable to a chemical separation of the elements. Distinction must be made here between methods suitable for: (1) samples of a small size such as minerals, rare substances, products of special separation procedures, concentrates, or simply rocks where one wants to avoid the introduction of contaminants by the bulk grinding media, and (2) bulk samples to be crushed and ground in considerable quantities for such purposes as combined major and trace element analysis, geochemical studies, statistical investigations, preparation of standards, or any samples taken for commercial purposes.

Samples are washed with water and if greasy or oily with alcohol or ether. In general, a careful examination of the pieces to be crushed must be made to eliminate possible external contamination resulting from transport and original sampling. This is discussed in more detail below. Sometimes inclusions are noticed that had not been taken into account and must be removed. Before proceeding, samples must be air dried, powdered samples are naturally not washed.

Small samples of 1–2 cm are best crushed in hardened-steel mortars of suitable size by pounding, avoiding entirely the twisting or turning motion of the pestle. It is clear that losses in small samples by splintering should be avoided. Further diminution is best performed in agate, mullite, or alumina mortars by hand or in an automatic grinder adapted to these.

For general classical analysis the powder should pass at least a 100- or preferably a 150–200-mesh screen, but grinding to −400 mesh facilitates fluxing and attack with hydrofluoric acid. For X-ray spectrography grinding to pass a 325–400-mesh screen (−325 to −400-mesh powder) is important. For fast-neutron activation the grain size matters little. If ferrous iron (Fe^{2+}) determination is to be performed, one should not grind finer than −100 mesh and should perform the determination as soon after the grinding as possible. In dry desert atmosphere, vibration grinding to −325 or −400 mesh and immediate HF attack is recommended.

Large bulk samples must be ground mechanically and heavy contamination is difficult to avoid. The author has used hardened-steel plates for one half of a sample and ceramic grinding plates for the other half to determine impurities introduced by two types of grinding and thereby obtain better estimates of the true composition of samples. As 99%- and even 99.9%-Al_2O_3 grinding plates, vials, and balls are now becoming available from the Coors Porcelain Co., their use is recommended in this *dual grinding method*. The 85%-Al_2O_3 plates commonly used do present contamination problems, introducing in addition to Al also Si, Ba, Ca, and Mg.

Crushing of large rock pieces is performed on a hardened-steel plate with a heavy hammer. Exceptionally tough pieces are broken in a hydraulic rock crusher.

Fist-sized and smaller rock pieces can be crushed to approxi-

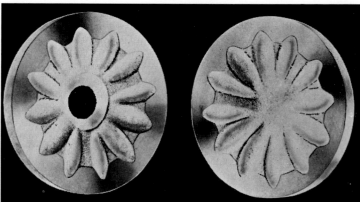

Fig.1. Bico-Braun pulverizer, Model UD and ceramic grinding plates. In the Nevada Mining Analytical Laboratory, Nev. (U.S.A.) two UD pulverizers are used. One is equipped with ceramic plates, the other with hardened-steel plates. Dual grinding and dual analysis then permit corrections for contamination of trace elements and the estimation of bulk contamination which has a dilution effect on the material ground. (Courtesy of Bico Inc.)

mately 1–2 cm size by passing through a mechanical *jaw crusher* of the Bico-Braun "chipmunk" type for example. The next step is to pass this material through a rollercrusher where the grain size is

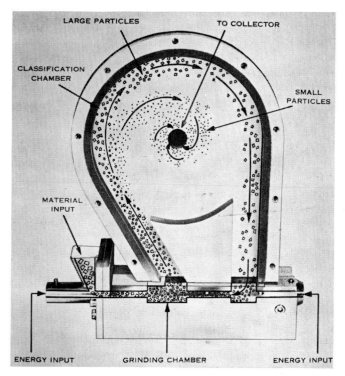

Fig.2. Operating principle of a Gem-Jet mill. Collision between particles driven by an air jet stream in the grinding chamber and their size classification in the main chamber produce a relatively uniform grain size. The whole operation can be controlled by regulating the air jet coming from the compressor. (Courtesy of Helme Products, Inc.)

further reduced to about 1–2 mm. The whole sample consisting usually of 3–4 kg is now split in a riffle into two equal parts, one to be ground entirely by hardened-steel plates, the other to be ground entirely by ceramic plates.

Grinding or pulverizing can be performed by "Laboratory disc

Fig.3. Ceramic-vial mills in a paint shaker for simultaneous grinding of several samples. Three ceramic balls are usually used as grinding media inside the vials. (Courtesy of Coors Porcelain Co.)

Mills" manufactured by Siebtechnik G.M.B.H. or in rotating-plate "Braun" grinders manufactured by the Bico Inc. (Fig.1), or in the "Trost Fluid Energy" mills by causing the particles to collide in a jet stream (Fig.2). The latter mills are manufactured by the George W. Helme Co., and seem to introduce little contamination. Powders so obtained should pass the 100-mesh screen.

If many samples are to be ground simultaneously, one can also use cylindrical ceramic containers with ceramic balls placed in series on regular paint shakers as done by the U.S. Geological Survey and the Spectro-Chemical Laboratory of the Coors Porce·lain Co. (Fig.3). The ceramic grinding vials and balls are manufactured by the Coors Porcelain Co. Particle sizes are given in μ and mesh sizes in Table I.

Fig.4. Plastic laboratory blender made of lucite. Standard model P-K Twin Shell, 8 quarts capacity. The base can be adjusted to accept two 1-pint (about $\frac{1}{2}$ l) or two 1-quart (about 1 l) shells suitable for smaller samples. (Courtesy of Patterson Kelley Co.)

The coarse powders (-100 mesh) are homogenized in V-shaped "twin shell" homogenizers made by the Patterson Kelley Co. (Fig. 4), stored, or further ground in the vibration-type "Wiggle-bug" or "Pica" cylindrical mills using large enough samples to avoid the breakage of the sleeves. The time of final grinding and the intensity

TABLE I

APPROXIMATE PARTICLE SIZES IN MICRONS AND MESH-EQUIVALENTS

Size (μ)	500	300	150	75	45	35	25
U.S. Sieve no. (mesh)	30	50	100	200	325	400	600

of vibration is adjusted so as to produce a powder 95% of which passes a 325- or better a 400-mesh screen. This grain size insures stable relative intensity for all elements in the X-ray-fluorescence method and is on the safe side when representative, small 5–10-mg samples are to be taken for optical emission spectrographic work. The grinding is performed in ceramic sleeves using ceramic balls for the one sample half and with hardened-steel sleeves and balls for the other sample half ground correspondingly.

The general scheme is:

(1) Sample, 3 kg → (2) hand crushing, hammer, splitter → (3) jaw crusher → (4) steel rolling → (5) riffle, division into two equal parts →

(6) ⎰ → (a) 1.5 kg, rotating plates, steel
⎱ → (b) 1.5 kg, rotating plates, ceramic   → (7) V-shaped homogenizers

(8) ⎰ → (a) 15 g, vibration mill, steel
⎱ → (b) 15 g, vibration mill, ceramic

The approximate grain sizes corresponding to the steps in the above scheme are: *Step 1:* washed and air-dried rock chunks; *step 2:* fist-size and smaller pieces (in geochemical work removal of the fine powder at this stage is still permissible, and is of advantage because this guards against incorporation of occasional larger splinters of metal in the rock sample); *step 3:* 1–2-cm pieces, and fines; *step 4:* grains < 0.2 cm, and fines; *step 5:* division into equal representative parts; *step 6:* about −100 mesh; *step 7:* homogenization, storage; *step 8:* 95% of −325 or −400-mesh particles.

This diminution and dual grinding method has been used by the Nevada Mining Analytical Laboratory for 8 years, and considerable analytical data have been gathered on the samples so obtained.

Contamination of samples during preparation

When solid samples come into contact with the grinding or

crushing media, mutual abrasion, chipping, or smearing occurs to at least some degree. This means that a certain degree of contamination during diminution is unavoidable. Foreign matter is often also introduced at the very source of the sample. Unclean surfaces may contain crusts of clay and salts deposited recently; paint, metallic stains, and rubber-tire marks have become too common. Putting samples into unclean bags that have previously contained material of different composition, like rocks into old ore sampling bags for example, or transportation of samples in tin boxes or cans, or rolling loose in the back of a truck, rubbing on metal, paint, rubber tires impregnated with antimony sulfide, all this may cause serious contamination in case trace elements are later to be determined. Therefore it is imperative to wash samples before crushing, even in alcohol or ether, if greasy, and scrub off the adhering rubber with a hard nylon brush, if necessary.

Pounding a rock with a hammer, similar action with a steel pestle in a steel mortar, or even jaw crushing or hardened-steel rolling does not constitute a major source of contamination as long as twisting or turning motions are avoided. It is the grinding with rotating plates that is the most serious cause for major contamination of the sample by the material. Recently the introduction of special cylindrical grinders and especially of air-jet impact grinders with teflon lining has reduced the contamination, but nevertheless a contamination-free grinding is still a dream of the analyst and geochemist. It may soon be that powerful ultrasound-shattering will reduce rocks and other solids to fine powders without any contamination. At present, however, the analyst has to cope with unavoidable contamination.

One possible solution to this problem has long been recognized. This is to achieve the diminution of separate portions of the sample with grinding media of different composition. The development of tough and hard mullite or ceramic plates for mills conventionally equipped with hardened-steel plates makes it feasible to divide a sample at a reasonably uncontaminated stage of diminution into two parts. The speed of the X-ray emission methods then makes a dual analysis possible. Using classical methods in such a procedure would obviously double the already tedious work and make the whole attempt impractical from a rational analyst's viewpoint.

When the dual grinding is used it is very important to select grinding material of a simple composition. The hardened-steel plates, for example, introduce mainly iron, but also traces of manganese, vanadium, and titanium and these contaminants appear when these trace elements are determined spectrographically. The ceramic plates manufactured by Coors Porcelain Co., were found to contain mainly alumina and silica, but also barium, calcium, and magnesium as major constituents. These plates also seem to introduce traces of Rb, Zn, Zr, La, and Ce. Different mullite and ceramic grinding materials vary in composition so that for each investigation it is necessary to determine the specific contamination.

One could assume that once the bulk contamination with a certain type of rock is established further double checks would be unnecessary. This is definitely not so, because many variable factors control the degree of contamination. In rocks the grain size, the intergrowth pattern of mineral grains, the texture, and the original grain size distribution affect the degree of abrasion of the plates. The speed of rotation, the distance of the plates which is variable, and the temperature changes may cause variations in contamination introduced by the grinding media.

The bulk contamination of a rock sample, by the combined methods described here, varies. In general, one may expect at least an introduction of a few tenths of 1% to nearly 1.5% of foreign matter when using the ceramic plates. Hardened-steel plates introduce from 0.1% to nearly 1% of metallic iron. No such contamination occurs if samples are handground in agate mortars.

In addition to problems that contamination causes by introduction of foreign matter, the *diluting effect* of the bulk introduced has to be taken into account in accurate analysis of major constituents. To demonstrate this effect it is sufficient to assume that a rock originally containing 50% SiO_2 has been contaminated by 1% of aluminum oxide (Al_2O_3) powder. It is obvious that, when analyzed for silica, this contaminated powder will yield approximately

0.5% less SiO_2 than that present in the undecomposed rock originally. If the ceramic plates contained silica, the calculation becomes more complicated. In the case of trace element determination, the dilution effect may be ignored, but not the specific
contamination by the same element introduced by the grinding
medium.

Other problems arise which have to be coped with in classical
analysis. The introduction of metal such as iron causes reduction
of ferric iron when the sample powder is dissolved in hydrofluoric
and sulfuric acids, giving raise to low results in reporting Fe_2O_3, or
high results for FeO. Removal of iron by a magnet does, however,
also remove one part of the magnetic minerals, magnetite for example, and, therefore, can not be applied.

All these problems are best solved by accepting the varying
degrees of contamination, but grinding by the dual method and
interpreting the results by double analysis of the same constituents
using fast instrumental methods. These methods are generally based
on the analysis of at least two to four different samples and the
double method also permits detection of gross errors in sample
preparation or weaknesses in the analytical methods.

Before concluding this chapter it may be emphasized that the
purpose of it is to orient and give one simple example of sample
diminution which copes with as many problems of contamination
as feasible. It is not the purpose of this book to give a compendium
of several useful methods of sample preparation.

Storage and care of samples

Well prepared samples are valuable references and contain the
final proof of work well done. Proper labeling and the use of glass
or suitable plastic containers well stoppered in such a fashion that
no contamination will result from long storage, are of great importance. One should avoid especially metallic screw-top containers
paraffin-saturated paper or cartons, and old rubber. Polyethylene
and other modern plastics are the best materials for storage of
samples.

Originally homogenized samples tend to segregate upon transportation and storage, even when moved from shelf to shelf. Larger particles and micas concentrate on the top of the powder, and heavier minerals or particles settle to the bottom. Fine powders as used especially for X-ray and spectrographic work tend to adsorb moisture, especially in humid atmospheres. One should correct for such contingencies when the powder is retrieved again for additional analysis or checks. The best way is to homogenize everytime before a sample is taken for analysis by tumbling in a V-shaped homogenizer for at least 10 min, and the first determination should be that of moisture or "minus water" as it is called by rock analysts. This check will reveal usually major changes of the sample during storage.

Ferrous iron may oxidize during long storage and chemical reactions take place if the sample contained acids, sulfates, sulfides, or metals. Samples believed succeptible to such reactions should be examined to determine if these reactions have occured, before undertaking further analyses.

In general a dry room with relatively stable temperature and low humidity should be selected for storage. One should avoid using this room for any grinding, crushing, or sample handling, unless performed under cover. It has to be a clean room with a minimum of dust and the samples on the shelves should be preferably covered by sheets of plastic which prevents contamination from fallen dust when the samples are re-opened. Unfortunately, these necessary precautions are sometimes entirely ignored and many storage rooms look like garbage dumps with half opened cans, spilled samples, old ore bags, discarded antiquated equipment, and even leaking acid bottles.

A geochemist as well as any conscientious industrial worker must guard continuously against sloppy handling of the sample material which has been obtained with such an expenditure of time, effort, money, and care. He must remember that this is the very basis and source of all the analytical work, research, compilations of data, conclusions, decision making of the mining geologist, and finally the ultimate proof of the geochemist's hypotheses and theories.

DECOMPOSITION OF SAMPLES

Enough powdered sample is generally available to the analyst so that preliminary tests may be conducted to establish the easiest decomposition method. However, unless the sample is water-soluble, the simplest attack is not always the one to be preferred analytically. The approximate composition and the whole sequence of the necessary determinations should be considered before a proper dissolution method can be selected. Only then is the sample brought into solution through a procedure least likely to interfere at any stage of the analysis. This is especially true when a total analysis of complex compounds is to be performed.

For example, most fluorine-bearing phosphates are soluble in strong acids, but if fluorine is to be determined, acid attack is not permissible because fluorine would partly volatilize as hydrogen fluoride. If silicates are present, some silicon in the form of silicon tetrafluoride would escape, rendering the determination of total silica invalid. Some compounds are made more soluble by ignition and this would seem a logical approach, but ignition may cause partial or complete loss of carbon dioxide, water, silicofluoride, mercury, sulfur trioxide and with it phosphorus pentoxide, thus either invalidating the analysis or causing some determinations to be incorrectly reported. Acid-soluble borates can not be dissolved in strong acids because of the volatility of the boric acid. Specific decomposition methods are discussed below.

In classical chemical analysis one has to plan and establish the analytical scheme according to the composition of the sample. Paradoxically this means that the composition of the sample has to be known (to some extent at least) before it is analyzed. Every analysis starts with a decomposition of the sample and the selection of the proper method is of utmost importance. It may save the analyst much valuable time if the sample is attacked properly from the beginning.

Undecomposed sample powders and their use

An ideal situation would be to analyze the rock sample undecomposed. This is only possible when high-energy electromagnetic excitation or neutron activation is used, and even then most rocks or polished rock and metal surfaces yield poor results due to unequal distribution of composition in different phases or mineral grains. Only fine-grained equigranular rocks may be analyzed with satisfactory results in this manner.

Representative sample powders can be analyzed very well by electromagnetic excitation with X-rays or by neutron activation. For the X-ray analysis the powders are preferably pressed into pellets and the smooth surface irradiated. For neutron activation the samples are packed in plastic containers, the so-called "rabbits", and irradiated. Both of these methods are uniquely suitable for the nondestructive instrumental analysis of sample powders giving excellent results in terms of precision and assuming careful calibration, fully acceptable results in terms of accuracy.

The main advantage of these nondestructive methods lies in the fact that substances analyzed need not be changed physically nor chemically in order to obtain data on their composition. This is important in analysis of total composition where elements displaying multiple oxidation states, volatile elements, or elements easily introduced during the routine procedures have to be analyzed. Any chemical manipulation, including fluxing, will introduce or remove some unaccountable quantities of oxygen, making a determination thereof futile. Most reactions in a laboratory take place in an oxygen atmosphere, therefore, in the case of oxygen, the unique advantage of nondestructive techniques is obvious.

We have seen that any handling of a sample may introduce contamination or at least change the particle distribution. This is also true when the weighing is considered. Further, one must realize that decomposition by acids, bases, or fluxes of any kind will inadvertently introduce foreign matter or cause certain losses in the sample. This is to be kept in mind especially when any trace elements are to be later determined, hence the importance of blanks emphasized in the analytical section. Major amounts of such main constitutents as silica can be introduced in solutions of old ammonia if stored in glass bottles, or by evaporations of ammoniacal solutions in glass beakers.

Since the analyst is interested in the original composition of the substance, the principle of minimum handling and of the greatest possible simplicity of the method, applied whenever feasible, is one of the basic rules in sample treatment during the whole process of analysis. This principle is also statistically sound because the errors of all the independent analytical steps or manipulations are accumulative. Naturally, there are numerous accurate and precise methods which may involve complex and numerous manipulations requiring fluxing and dissolution of sample.

Solution techniques

Solution of sample in water[1], acids, acid mixtures or bases, and organic special solvents may be regarded as the simplest way to render it into a suitable form for chemical attack and separation. It has been emphasized above that the seemingly easiest way is not always the best way analytically. Major silicate rocks do not yield to simple acid attack except to that of the hydrofluoric acid. Most ore samples can be attacked by dilute or strong acids or their mixtures when only commercial-type analysis is desirable, because the insoluble silicates are of little interest to the mining man and can all be dumped under one label "silica" or "insoluble residue". This residue seldom carries the metals of main commercial importance and if it does, these elements usually are not recoverable by the methods used in the beneficiation of the ore, so that it may not even be desirable to report that part. These methods are of the leaching type and are unsuitable in meticulous geochemical investigations.

Specific solution methods are given in the text with analytical methods for the elements. It may be emphasized here, however, that during most stages of the classical silicate analysis sulfuric acid

---

[1] Distilled water is used in all procedures given in this book.

is undesirable, and whenever it is introduced or suspected, precautions must be taken for its complete removal before proceeding to the next stage.

In the selection of the type of acid to be used, one should consider the composition of the sample avoiding an acid that would cause a precipitate or insoluble residue to form. An acid that can be easily removed by evaporation is always preferred. Mixtures of acids are avoided unless they are of specific oxidizing characteristics, as aqua regia for example. When acids or acid mixtures do not dissolve or attack the sample the addition of oxidizing reagents such as bromine, or of strongly oxidizing acids such as perchloric acid may be helpful. Extreme care should be exercised whenever these compounds or hydrofluoric acid are used, because skin injuries difficult to heal and explosions are possible.

One always applies first dilute hot acids and allows time for decomposition. Only when this is unsuccessful should a stronger or concentrated acid or an acid mixture be tried. Dissolution whether effervescent or not is always performed under cover to avoid losses.

Roasting the sample if one fears no losses of volatile substances will often render it soluble to acids if one wants to avoid fluxing by all means. When all other acids fail, and one does not want to flux for some reason, hydrofluoric acid with a few drops of concentrated sulfuric or perchloric acid may be used in a platinum dish or teflon beaker. Thus silica is removed and lost analytically and the precipitation of the insoluble calcium fluoride and barium sulfate may cause difficulties, but nearly all samples are decomposed.

This attack gains special significance in the so-called *Berzelius method* for the determination of the alkali metals, and is used today mainly in preparation of solutions for flame-photometric determinations of alkalies, and the alkaline earths, the main problem being the partial precipitation of calcium as fluoride, which has to be taken into account in order not to lose part of this element when filtering the insolubles. The use of perchloric acid instead of sulfuric may in this case be especially recommended. In some laboratories there is a trend to replace sulfuric acid by perchloric as far as possible.

Fluxing techniques

The beginner may assume incorrectly that unless silica is to be determined, the hydrofluoric acid attack would be the best method in most cases to decompose silicates. Disregarding the importance of fluxing and trying to bypass it by all means would be a most serious error. Fluxing is as old a procedure for the initial decomposition of samples as analytical chemistry is a science. Berzelius developed fluxing into a manyfold skill which still today forms the first attribute of a successful analyst. Fluxing is not only a means of decomposition, it is also often a very valuable form of efficient separation of elements. Nevertheless it is also true that over-emphasis on fluxing because of its historically entrenched position, may lead astray a chemist who uses otherwise modern methods.

A geochemist usually compares the compositional differences of rocks more or less similar in chemical as well as in physical properties, and thus a truly nondestructive properly standardized technique which gives relative intensities for most elements is an ideal approach for him. A cement analyst analyses similar materials, the final product or the raw materials, and compares them to certain desirable standards. He does not compare the composition of the cement powder to that of a limestone for example. In these and similar cases, the fluxing would seem to be of a definite disadvantage since it in itself causes contamination, dilution, hygroscopicity problems, and necessitates several weighings. Also, a possibility of escape of volatile substances has to be considered.

Only a few generally applicable fluxing reagents can be mentioned here. Generally one fluxes with anhydrous *sodium carbonate*, Na_2CO_3, or its mixture with potassium carbonate, K_2CO_3. In classical analysis of silicates, phosphates, and fluorides, or when such anions as sulfate, phosphate, chlorine in silicates, fluorine, chromate, arsenate, molybdate, antimonate, tungstate, niobate, or tantalate are to be determined, this flux is ideal. It has the advantage that it is also an excellent separating agent for such commonly determined elements as Ca, Mg, Ba, Sr, Fe, and Al. When the cake is leached with hot water, these cations remain insoluble as carbonates. The above mentioned anions remain, however, in solution. Often this flux turns out to be the fastest and most efficient separation method, even though at the beginning it may seem somewhat cumbersome.

Another frequently used fluxing material is *potassium bisulfate*, $KHSO_4$, or the "lowest temperature flux". This is used mostly to decompose the ignited aluminum and iron oxides separated during the analysis. For silicate rocks this flux is less desirable and less effective than the carbonate flux. A platinum crucible or even Vycor or Pyrex tube can be used, but a sintered silica crucible is recommended, and more economical.

Sodium peroxide, Na_2O_2, is a very powerful flux for most substances not yielding to the attack of other fluxes completely, but its effectiveness is such that no suitable vessels are known that would resist its attack sufficiently. One must work with nickel or iron crucibles which are strongly attacked. A specific case where the use of this reagent leads to a separation as well as to an efficient decomposition is in the analysis of chromium ores and chromium-bearing refractories. Platinum can not be used, unless special precautions are applied, which constitute considerable complication of the whole process and the crucible may still be damaged.

Boric anhydride, B_2O_3, or *borax*, $Na_2B_4O_7 \cdot 10H_2O$, are powerful fluxes but since boron must be removed after the decomposition by methanol treatment, their use in practical analysis is restricted to special cases, such as chromite and zircon. Alkali-free boric oxide may be used to advantage if the sample is small and alkalies as well as other constituents have to be determined. Platinum can be used.

Fluxing techniques for sodium carbonate, potassium carbonate, borax, peroxide, and boric acid consist generally of the following *procedure:*

Mix the ground, dry sample well with the flux, preferably with mortar and pestle, in approximate proportion of 1/3–1/5. The importance of using the finest attainable sample powders can not be over-emphasized here. Cover the crucible bottom with a layer of pure flux powder, transfer the mixed sample into the crucible, "wash" the container with flux powder into the crucible, and cover the main portion of the sample with 1–2 g of pure flux to prevent losses during initial stages of heating. The 20–30-ml platinum crucible should only be one-third full. Cover the crucible incompletely with the lid, leaving a narrow passage for the escaping gases. Heat first gently and increase the gas flame gradually, attending the process closely all the time and decreasing the heat if the contents start raising and spattering. When the contents melt, gradually increase the flame to full blast observing intermittently whether any particles remain on the crucible bottom. As soon as the melt appears clear, discontinue the heating, swirl the contents carefully so that most of the solidifying melt sticks to the walls of the crucible, and cool for 1–3 min. Place the still hot crucible into distilled water in a covered beaker to ease the disintegration of the melt by induced cracking. Proceed further as advised in Chapter 7, under section on silicon. Note that the borate melts are very viscous and that the peroxide melt is not transparent.

Fluxing with *potassium bisulfate* differs insofar as the sample powder is put directly into the crucible over some bisulfate, covered with bisulfate, and if desired, mixed. The deliquescent nature of bisulfate prevents efficient mixing beforehand and causes the substance to stick to the walls of the mortar and pestle, or a mixing rod. Special care should be exercised during the initial stages of this fluxing technique because it tends to boil over if over-heated. These difficulties can be avoided if pyrosulfate, $K_2S_2O_7$, is used as a powder instead. It can be prepared by melting and boiling the bisulfate till effervescence ceases, cooling and grinding the cake.

Special fluxing techniques

Calcium carbonate–ammonium chloride fusion, as described in Chapter 7, is the most frequently used special sintering technique in analysis of rocks, minerals, cements, slags, and glasses. It is called the *J. Lawrence Smith method* and it is interesting to note that Berzelius, who otherwise was a master in the use of fluxes, used in 1824 the hydrofluoric-sulfuric acid attack, and that Smith developed his method almost fifty years later. The main advantages in using this method are the exclusive separation of alkali metals from all other constituents in a complex silicate mixture, and especially the retention of all magnesium in the water-insoluble residue, which simplifies the determination considerably. The selectivity of separation of alkalies by this method is such that it is often advisable to use it even if the sample is soluble in mineral acids other than hydrofluoric. In routine *flame-photometric analysis*, however, when tens or hundreds of samples are decomposed simultaneously, the *acid decomposition* becomes more efficient.

Other special fluxes, mainly with some added oxidizing or stronger caustic properties are used. Addition of some *potassium nitrate*, KNO_3, to the sodium carbonate helps to oxidize sulfides to sulfates, arsenic compounds to arsenates, and chromium in its minerals to chromates. This procedure requires skill, and great care should be taken not to damage the platinum crucible by excessive addition of the nitrate.

Another good flux is *lithium carbonate*, Li_2CO_3.

Potassium hydrofluoride, KHF_2, is a useful flux for zircon, titanates, tantalates and niobates, and some refractory materials. Platinum crucibles must be used. The advantages are the low temperature of fluxing and the short time required. It is, however, possible to bring these materials into solution also with hydrofluoric-sulfuric or perchloric acid mixtures.

Contamination of samples during decomposition

Contamination by abrasion during diminution is not the only source of introduction of foreign matter to the sample. The addition of major quantities of so-called "chemically pure" reagents is often the source of impurities. Fluxing for X-ray fluorescence may introduce unknown amounts of trace elements to the fused samples. Acids, bases, and fluxing media in general are commonly used in such amounts that they must be continuously suspect of the introduction of different elements. Even impure gas flames and carbon dioxide in the atmosphere have to be recognized as possible sources of contamination, as pointed out in Chapter 7 under the section on sulfur. To list all possible offenders that may escape the attention of a beginner would require more space than is feasible here, but the main contamination sources are mentioned in the analytical part of this book. The analyst is advised to run complete blanks and check for impurities as often as possible in accurate work.

Storage of decomposed samples

Decomposed samples are attacked chemically to separate the elements. Often the total sample is used up in this process. In

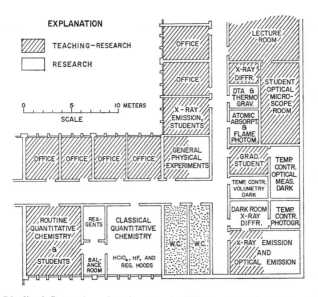

Fig.5. Idealized floor plans for the Nevada Mining Analytical Laboratory, Mackay School of Mines, University of Nevada. The figure shows the division of research, teaching, and routine type work, which overlap where applicable. Note the temperature controlled research rooms for volumetric, optical, and photographic work. Note the different exhaust hoods necessary for silicate analysis and safety. The reagents and standards are accessible only to research personnel. Once opened they are used exclusively on either the routine or the research laboratory shelves.

many cases, however, the dissolved sample is diluted to a specific volume and aliquots of this liquid are used for the different determinations. In this case it is of advantage to have a constant temperature storage or "volumetric room" where all liquids are kept well stoppered. Such a room has been constructed in the Nevada Mining Analytical Laboratory, as shown in Fig.5.

The storage of pulverized decomposed fluxes for X-ray analysis presents also some problems considering that many fluxes are hygroscopic. It is, therefore, necessary to keep the powders in tightly closed bottles. In general, the so-called "water-free" fluxing media are often apt to adsorb some surface moisture after fusion, especially when pulverized. Fine powders take up more water than coarse grained materials.

CHAPTER 3

Statistical Evaluation and Sampling

Any measurement is more or less liable to error, and most values found will vary from observation to observation. This fluctuation of data indicates the presence of errors. Thus, fundamentally, an analyst wants to estimate the inherent error in his work. He has to evaluate his data statistically in order to know the quality of the work performed and to represent the results properly. This requirement has to be met in connection with the first steps of sampling and during every subsequent treatment of the sample. Statistical interpretation of results permits an unbiased presentation and the determination of the degree of usefullness of data collected. It enables the chemist or geochemist to express in mathematical terms the reproducibility of results, called *precision*, and to evaluate the measure of correctness of the data, called *accuracy*. Due to the scope of this book, only a very incomplete introduction to statistical evaluation can be given.

Prior to the reporting of analytical results, tests for precision have to be performed. Every new method, and even partial changes in established procedures, or the acquisition of new equipment, require careful tests of precision before any data can be reported. Periodic checks to assure the continuity of the once established precision are mandatory. When prepared samples of known composition or well analyzed samples are used as standards for these purposes, the analyst also learns how closely his results fall to the "true" values. Performing or planning a series of complex experiments to be interpreted statistically later, an analyst is well advised to consult an experienced statistician beforehand.

Failure to recognize the difference in the meaning of the terms "precision" and "accuracy" leads to serious misunderstandings.

Not quite aware of this distinction, an analyst implies sometimes quite unintentionally that his data are accurate because they are precise. The rapid spread of instrumental methods of elemental analysis, which are inherently precise but inaccurate unless standardization is performed properly, has lead to premature claims of superiority of certain new methods. One may, therefore, be reminded here that the same error repeated over and over again can lead to perfectly reproducible results.

Two types of error can be distinguished. The *random* error originates from irregular fluctuation of single results, meaning that figures will vary irregularly. The causes for this may be due to vibrations of instruments, temperature variation, power surges, errors in reading the scale, etc. None of these causes can be completely controlled because of their irregular fluctuating nature, nor can their effects be accurately predicted or measured. The *systematic* error is due to bias in sampling or method, meaning that most figures will tend to be either higher or lower than the correct composition. It is intrinsic to the method of observation and thus has a constant effect as long as a certain procedure is used. When actual results are interpreted statistically both these errors appear superimposed.

In statistics the random error or *precision* may be expressed in such mathematical terms as:

$d = deviation$ of a specific result or observation from the arithmetic mean, $\bar{x}$;

$S = \sqrt{\Sigma d^2/(n-1)} = standard\ deviation$, where n represents the number of determinations; this expression provides us with a measure of the uncertainty of a single observation based on the total number of observations made on a sample; S is expressed in the same units as the original data;

$V = \Sigma d^2/(n-1) = variance$, or the mean square deviation, which gives us an estimate of the spread of results from the mean: obviously we can write $V = S^2$;

$C = (S \times 100)/\bar{x} = relative\ standard\ deviation$, or *coefficient of variation*, which tells us the uncertainty of a single observation in percent, in terms of the total relative quantity of the constituent in the sample. For the practical analyst this value is important because it permits the evaluation of the *relative error*, $E = C/\sqrt{n}$ in groups of different samples as well as the evaluation of the methods used. The knowledge of the deviation from the mean value leads thus to an ability to evaluate accuracy. This helps to assure data reasonably close to the true value. An example of evaluation of analytical data is given on p. 243.

In an ideal case, when no bias exists in any of the sample retrieving, preparation, or analytical methods, accuracy can be regarded as synonymous to precision. This is often the assumption of the chemist reporting his results. Accurate results can be best assured when high precision is combined with unbiased methods.

The introduction into practical problems of statistical interpretation of sampling and analytical data may best be done by grouping some data on samples and showing what happens when they are represented graphically. A geochemical sample is usually part of a much larger body of rock, ore, water, or gas. (Statistically a sample may also be a single result from within a series of determinations.) One speaks of the *population*, which denotes the total quantity of the body represented by the sample. The population is an aggregate of all conceivable observations on some specific characteristic or a random variable. When results are grouped according to selected ranges and plotted, a histogram, which shows a certain *distribution* or frequency of data, is obtained. If the number of intervals is increased, an infinite amount of ranges could be finally assumed and this histogram would become a smooth curve (Fig.6). This differential curve can also be represented as an integral by adding up the data consecutively and plotting in the manner indicated in this figure. Most unbiased sampling procedures of homogeneous populations as well as plotting of data of chemical analyses yield approximately symmetrical distribution curves. These, when idealized, are called *normal* or *Gaussian* (Fig. 6). Accordingly, the statistician speaks of normal, or Gaussian distribution. A family of normal distribution curves is shown in Fig. 7.

One must realize that the normal distribution curve is only an approximation that happens to provide a convenient means of

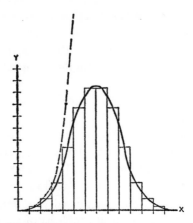

Fig.6. A histogram with corresponding normal or Gaussian distribution curve. A corresponding integral curve is indicated in part by a dashed line. Disproportional size relationship makes it impractical to present the whole integral curve on the same scale. Actual histograms composed from analytical results rarely show the perfect symmetrical relationship represented in this figure and are usually somewhat skewed on either side.

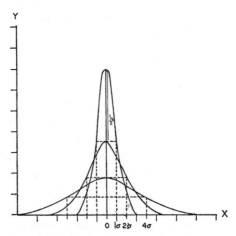

Fig.7. A family of normal distribution curves corresponding to standard deviation of 1, 2, and 4 units. The vertical line is hypothetical for absolute precision. Graphic determination of standard deviation is done by drawing a horizontal line at the half-height of the curve and dropping a vertical line toward the x-axis from the point where this line touches the curve.

description of collected data. By a closer analysis of the shape of actual distribution curves one discovers generally some degree of skewness. Mixed distributions showing two or more maxima are also common. Thorough knowledge of the sampling and the analytical method may then help find an explanation in a form of a built-in bias in the methods used. Thus an analysis of negative and positive skewness of chemical results on rocks W-1 and G-1 (STEVENS and NILES, 1960, pp.19–30) has helped to point out certain weaknesses of the methods used.

The *spread* characterizes the precision of the method. Higher precision means a sharper curve, which is demonstrated by the extreme, hypothetical case where the standard deviation is zero. This would mean that 100% of all data should fall in this infinitely narrow range. When the broader curves are considered we notice another extreme but real possibility. It becomes clear from the asymptotic nature of the tails (which approach infinitely close to the x-axis, yet never touch it), that an improbable observation can only occur with infinitesimally small probability. Although an analyst realizes this fact, he is nevertheless always surprised when this happens.

The areas under the curves (Fig.7) can be considered as representative of the entire population which is being investigated. If we select a symmetrical portion of the area corresponding to 95%, and if the total amount of determinations performed was hundred, ninety-five determinations, or nineteen in twenty, should fall within this range, and one result out of the twenty should fall outside the selected limits. One speaks in this case of *probability* of 5%, denoted often: $P = 0.05$. A probability of 0.50 means a "fifty-fifty" chance, which is merely a "yes"- or "no"-proposition.

Standard deviation can be used as a measure to limit the accepted data on both sides of the mean. Greek sigma (σ) is used as notation for standard deviation of a *population*. Drawing vertical lines (Fig.7), and calculating the corresponding area of the normal distribution curve results in the following approximate values:

$\pm 1\sigma \cong 66\%$ of population included; $\pm 2\sigma \cong 95\%$ of population included; $\pm 3\sigma \cong 99\%$ of population included.

The statistician usually speaks of *confidence limits* in percentage figures of 66, 95, and 99%, which corresponds closely to one, two, or three standard deviations on both sides of the mean value.

In general sampling or chemical practice the precision is reported with confidence limits of $\pm 1\sigma$, or $\pm 2\sigma$. Too often, indeed, an analyst makes no attempt to appraise the precision of the method in his own hands.

In an ideal unbiased case the accuracy of observations can be assumed to be identical with the precision. Thus taking the usual confidence limit of two standard deviations we may write $A = 2\sigma = = 2\sqrt{V}$, where A is accuracy, and V is variance. Variance is convenient to use as an indicator of the magnitude of an error because variances due to different independent random errors may be added to represent the total error. In our case errors or variance introduced by sampling, V_s, by diminution and decomposition, V_d, and by the analytical techniques, V_a, may be considered. Then the total variance is:

$$V = V_s + V_d + V_a \tag{1}$$

and if no bias is present:

$$A = 2\sqrt{(V_s + V_d + V_a)} \tag{2}$$

This equation permits us to calculate the precision necessary in an individual sample handling or analytical process to estimate the accuracy required.

For example, we get for sampling precision or variance which allows for decomposition and analytical errors:

$$V_s = A^2/4 - V_d - V_a \tag{3}$$

This equation has special significance for the geochemist attempting to establish sampling procedures to a specified accuracy with a predetermined probability.

An analyst is always ultimately interested in accuracy, yet when queried about what accuracy he has exactly in mind, a scientist analyst realizes often that this requirement is a rather arbitrary measure, which depends largely on convention, on what others had reported or claimed, or on a general desire to do the work as well as one can. In industrial chemistry and metallurgy, specified standard requirements facilitate the exact answer to this question.

The next question is, how can a specified accuracy be achieved? Obviously the proper design and control of sampling forms the crucial step or basis for all representative data to be obtained in the future. Thus one is confronted with (a) the variability and size of the population which may be a rock body, a pile of ore, or slag, a body of water, bags or a small bag of powder, etc.; (b) the weight of a sample to be taken and the number of such samples; (c) the spacing of such samples over the whole body; (d) and finally, the diminution processes to be used to produce the final representative sample.

From this list we see that the formula above greatly over-simplifies the actual conditions and that many factors influencing the precision can be recognized within the general categories of sampling, sample preparation, and methods. For example, several random small samples give better accuracy than single large samples when the material is variable, such for example as a migmatite or coarse porphyritic rock. Systematic errors or bias may invalidate any careful sampling or analytical procedure and are often very difficult to detect because, when grouped, a bias will affect the positions of the distribution curves giving biased means.

Random sampling to represent the whole population is done by numbering the sample location or containers and selecting at random a pre-determined part of the numbers using *Tables of Random Numbers* provided by several texts on statistics. This method can be refined by dividing the population into layers. These types of sampling are mainly adaptable to industrial materials and some rocks, and their main weakness is that one single sample taken from heterogeneous material may be far from representative.

In geochemical work, as well as when sampling industrial products and raw materials, the common practice is to divide the population with a grid and to take increments evenly spaced. This is called *systematic sampling* and it may be subject to bias, as mentioned (see p. 8). Not all samples taken need be analyzed nor all grid intersections be sampled. The random selection of spots to be sampled or samples to be analyzed may be done with the random tables or by mixing all the numbers written on identical pieces of paper and drawing a certain pre-determined part of them as in a lottery.

After selecting the sampling locations randomly or spacing the collection sites systematically the problem of the sample size and the number of samples or increments arises. A practical description of sampling methods for ores and industrial minerals also applicable to rocks is given by TAGGART (1945).

A *sample size* sufficient to characterize the sample site but not too large for convenient handling is an ideal compromise. As a practical estimate and applicable for major constituents one may take approximately a fifty-fold volume of the largest fragments or crystals present in the material. Thus a porphyritic granite with its largest feldspar phenocrysts of approximately 1 cm^3 would require a minimum of about 250 g of sample. Actually, whenever possible, larger samples are recommended as stated before (p. 16), and one has to consider that the sample size is also largely a question of available coring or other sampling facilities. Obviously all samples have to be of proportional or equal size to avoid the dichotomy that one sample is bigger and thus more representative than the other.

In geochemistry one has often to resort to over-sampling in terms of sample quantity and frequency. This is due to the often encountered uncertainty about the variability of the population studied. The *number* of samples to be taken is therefore difficult to estimate.

It is important to realize that variance is inversely proportional to the number of samples taken. To understand this the analyst has to recall that whenever he makes just a few determinations they usually agree quite well but as soon as the number of determinations increases, occasional larger deviations will usually occur. The probability that an error will occur, increases with increasing number of data. For example, when working with a 95% confidence limit one has performed twenty determinations it becomes highly probable that one determination will yield a result outside the accepted limits. This means that an error computed on the basis of a few determinations will generally appear to be smaller than that computed using a larger segment of the population. Yet, unless a sampling or purely analytical error is discovered, there is no reason to discard this sample since it also is a representative fraction of the total population.

All aspects of a geochemical or analytical problem must be considered in order to be able to interpret the results statistically. Discarding of any data prior to statistical analysis is premature unless a specific cause exists or when a gross error of known origin is detected, otherwise serious falsification of the quoted precision will result. In the first case an understanding of the goal is required in order to justify the rejection of any results. For example, if we are studying the variation of composition of a basalt flow as a part of a general study of the composition of unaltered basalts, we may be justified in discarding an analysis showing nearly 100% SiO_2 and probably indicating a quartz or chalcedony veinlet so common in these rocks. If, however, we were studying the inhomogeneity and the degree of hydrothermal alteration of a block of basalt in the geological sense, a siliceous veinlet would definitely represent a true part of the whole population, and no justification whatsoever would exist for its omission from the data collected.

We can generalize here and state that the number of samples depends on the problem. In a rigorous system only actual analysis and subsequent statistical interpretation can show whether variations in composition reported are significant in relation to the combination of possible errors due to sampling, sample preparation, and analytical methods used. One thus goes a complete circle in order to be able to answer this question. To determine the significance of errors one uses the *analysis of variance*, which attempts to examine separately the errors due to the methods, and those due to the nature of the sample or the factor studied. Experiments for such studies require a pre-established design before the sampling is commenced. The scope of this book permits us only to indicate the problem. The analyst, confronted with a difficult sampling or interpretation problem, will consult more comprehensive texts on this subject, listed in the Bibliography.

It was stated above that taking large samples from inhomogeneous populations defeats the purpose of sampling and results in cumbersome samples. Considered in terms of variance we may state that large variance in the population has more effect on the inaccuracy of results than repeated analyses of a single sample. There is often no need for repeated reproducible determinations of a factor in a sample taken from an inhomogeneous population because this could not significantly better the total variance. Single analyses of additional random samples will give more accurate results and will take less time.

In mathematical language we can describe the total variance, V_Σ, as consisting of the sum of two different variances, V_{rep} and V_{inhom}, based respectively on the reproducibility of analysis, and the inhomogeneity of the object studied:

$$V_\Sigma = V_{rep}/n + V_{inhom} \tag{4}$$

Obviously if V_{inhom} is considerably larger than V_{rep} a larger number (n) of analyses of a single sample could not have a significant effect on the total variance.

It is clear that the number (n) of analyses, samples, or any factors necessary to achieve a given accuracy (A) will depend on the actual variance (V) displayed by these factors. For practical purposes an approximate evaluation of this number can be made using the following equation:

$$n = 4V/A^2 \tag{5}$$

The statistical analysis of variance helps to answer questions concerning the relative contribution of different sources of error to the final results. To list a few problems that one may encounter we shall mention only the random errors, the systematic methodical errors, the possible personal biases of the chemist and even the bias due to the general conditions of a laboratory. Studies of intra-laboratory versus inter-laboratory precision can also be conducted using the analysis of variance. Investigations of this sort require much time consumed on calculations. Therefore, one uses electronic calculators and computers depending on the complexity of the problems. The individual steps are programmed for example as follows: (*1*) estimation of the mean; (*2*) calculations of the variance; (*3*) testing of the significance of differences between the mean values and analysis of variance; (*4*) correlation of different factors (elements in geochemistry) of a population; (*5*) construction of systems of trends and gradinets in the form of maps, enabling unbiased judgement by the geochemist.

CHAPTER 4

Factors Influencing the Selection of Methods of Separation and Analysis

The purpose of this chapter is to give a general outline of methods which may be suggested due to their relative simplicity.

With the revolutionary development of instrumental methods simultaneously with refined separation and analytical techniques, the beginner is often confused when forced to decide what methods or combinations of procedures to use. In this case a haphazard decision based more or less on the assumption that one analytical approach is as good as another, often may lead to a selection of excellent techniques totally unsuitable for the composition of the sample. Refined procedures are generally difficult to follow with the limited skill and training which are natural attributes of the young analyst, and a geochemist is often invading the field of analytical chemistry from the realm of the earth sciences and does not have at first the caution necessary to deal with strictly quantitative problems. A few general principles for selecting a method are therefore discussed in this chapter. When dealing with major constituents, specific problems of trace element analysis may be ignored here, but many of the general principles apply in both cases.

Evaluation of the means of the laboratory and the minimum requirements for performing the task satisfactorily is of primary importance.

Classical, well established, gravimetric methods can usually be adapted with the minimum of expenditures. In terms of initial minimum investment, only a balance, an oven, a hot plate, gas burners, some glass and porcelain ware, and a couple of platinum crucibles, and a few suitable reagents are required.

For restricted purposes a fully satisfactory classical type laboratory can be established for a few thousand dollars and operated at a cost of a few dollars per determination, depending, of course, on the total amount of work and determinations done. If only a few elements are of interest in a material of restricted physico–chemical properties, as for example copper, zinc, and lead in ores, an initial expenditure of about one thousand dollars will assure satisfactory "wet-chemical" industrial analysis.

If on the other hand a nondestructive instrumental analysis is attempted, the initial cost will be tens of thousands of dollars. Such an expenditure must be justified by the number of samples and the need for more complex and fast analyses of several constituents.

A satisfactory analysis of complex inorganic material can hardly be performed with one piece of equipment, but it is usually not necessary to combine all possible modern methods in a routine laboratory. In an ideal laboratory capable of accurate, fast, and complex analysis, one tries to combine a minimum of equipment and methods. Each method or piece of equipment has to be selected considering its specific characteristics favorable for analysis of an element or elements in the concentrations usually encountered. It is unwise from an analyst's standpoint to try to adapt a marginally suitable technique for routine analysis if better methods are accessible. Therefore, a certain degree of complexity in an analytical laboratory is unavoidable. In general, an analyst should work toward gaining the ability to compare any of the instrumental results with results obtained on the same material using conventional classical methods, preferably gravimetric. Most *standards* for major elements are based on gravimetric analysis.

SEPARATION OF ELEMENTS

After decomposition of the sample the main effort in classical analysis consists of separation of the elements into groups. These groups are further treated until specific elements are precipitated individually. In modern spectrographic, flame-photometric, atomic-absorption, and other spectral techniques, the relative intensity of the electromagnetic radiation is affected at least to some degree by other elements present, and although many methods to minimize this have been designed, in accurate work it is of advantage to separate the elements at least into chemically coherent groups, before the application of any of these techniques. This means that classical methods of separation, for example, separation into the ammonia group, the hydrogen sulfide group, and the alkalies, are still playing a major role in everyday-routine instrumental analysis. Because of the need of efficient separation of the elements before determination, even in such essentially nondestructive techniques as neutron activation, one of the main developments of analytical chemistry has been in the application of extraction, chromatographic, and ion exchange as well as co-precipitation methods.

In our case, the extraction and the ion exchange methods are of specific interest. Only a few general methods and practical hints may be given here (see p. 53). These partition methods have the advantage of giving simple, fast, quantitative separation of elements or groups of elements for further classical or instrumental treatment.

GRAVIMETRIC "ABSOLUTE" METHODS

Gravimetric methods are suitable especially for the accurate determination of major components in inorganic mixtures. Satisfactory methods have been developed for gravimetric determination of most metals and non-metals as well. The great versatility of these methods is remarkable. In some cases the classical gravimetric procedures are clearly superior in accuracy to all other methods, as in the case of silicon. Most standards available to the inorganic chemist are based mainly on gravimetric work.

The nature of these methods consists of the separation of a single oxide or compound of an element from a complex mixture. This permits further purification of this precipitate and a check for trace impurities, assuring results which are inherently near the "true" value. This is especially true when a total analysis, such as the silicate rock analysis, is performed. In that case care is taken to check and identify every consecutive precipitate so that if a certain amount of an element present escapes precipitation, or appears as an impurity, it is "caught" or eliminated during the next step.

Silicate rock analysis may serve as a good example of a relatively imprecise method yielding relatively accurate data. The time consumed and the effort necessary to conduct such complete analysis precludes the fast accumulation of a series of parallel observations on different increments of the samples. Thus, the basis for statistical evaluations is often lacking and the analyst must rely on painstaking care in checking the purity of his precipitates. When an experienced, cautious chemist reports the results of a single determination one is usually justified to accept the data as reasonably true. We notice here that one of the main handicaps of the classical gravimetric analysis is the long time necessary in order to produce good results. Nevertheless when only a few rare samples are to be analyzed or original analytical work of great importance is required, the classical methods will usually prove to give the most reliable basis for further instrumental work.

A modern laboratory employing mainly instrumental techniques may be very helpless when initial information on new substances is required, if it has not retained at least the bare essentials necessary to analyze by the classical methods. Disastrous consequences may result from relying solely on instrumental analysis and neglecting at the same time the needs and the necessary improvements in the "wet" chemist's corner, if indeed any such corner is left for him at all. Chemists pushing buttons on their highly sophisticated electronic monsters tend to forget that to transform the relative counts or intensities that they are measuring into accurate meaningful figures they must use data originating mostly from the bench of the classical gravimetric chemist.

Practical limits of quantitatively meaningful detection by gravimetry are in the neighborhood of one part in a hundred.

INSTRUMENTAL NONDESTRUCTIVE METHODS

When the speed of major element analysis and a multitude of analyses of samples of similar physicochemical properties is the main requirement, instrumental nondestructive techniques become competitive with gravimetric, volumetric and flame-photometric analysis. The analyst using these modern methods must realize that the accuracy of his results depends on working curves, which show the relation of the intensities measured to the amounts of elements in the sample. For all practical purposes this can be done only with the help of the "wet" chemist.

High initial cost of equipment is the main handicap of these methods. Their use requires good standards and specially trained personnel. Especially in the cement, mining, metallurgical, or similar industries, but also in geochemical research, where numerous samples have to be analyzed continuously, such investment usually pays off in a surprisingly short time and may even open new possibilities in industrial processes by providing "instant" data.

When faced with multiple analysis of complex samples for major as well as for trace elements, one may best select the X-*ray emission* as one of the most efficient methods applicable to concentration ranges usually of commercial interest. An addition of ultra-soft excitation, for example the Henke tube, may prove beneficial especially in the cement industry and for geochemical rock analysis, because it gives excellent results for such elements as Al, Si, Mg, Na, and S.

Where total oxygen is of special interest, as in some alloys, greases, oxides, sulfides or minerals in general, the addition of a *fast-neutron generator for neutron activation* work may prove to yield data practically unattainable by other methods. This author considers X-ray fluorescence with fast-neutron activation as the most useful combination of equipment suitable for total analysis of major as well as minor constituents in a complex inorganic mixture. There hardly exist two other basic analytical techniques which can provide accurate data with comparable speed and complement each other as well. Because most trace elements as well as all major elements can be determined by such a system, the principle of minimum equipment suitable for a maximum of specific purposes and also capable of performing efficiently a large quantity of work, is uniquely represented by this combination of equipment.

The lowest practical limits of semiquantitative detection by X-ray fluorescence are in the neighborhood of 1–10 p.p.m.

VOLUMETRIC METHODS

This branch of quantitative inorganic analysis which deals with such apparatus as the volumetric flask, the burette, and the pipette may be considered as one of the most important branches of analytical chemistry. *Titrimetric analysis* with its relative simplicity, speed, and very accurate results, has certain advantages over the gravimetric methods. For the analyst, speed means that precision of the method can be reasonably well determined and maintained at a high level in any laboratory without excessive effort. With measurable amounts of liquids and easily detectable color changes of solutions or indicators, reliable work can often be performed on the technician level once the method and conditions have been established. The ability to control closely the acidity and concentrations in liquids as well as sharp endpoints of the indicators, and the permissibility of back titrations assure accurate results as well. Several volumetric procedures are given in the text. One of the most important and unavoidable titrimetric procedures is the determination of ferrous and total iron by titration with permanganate or chromate. A quantitative laboratory should have at least a capability to perform permanganate, chromate, iodometric, and thiosulfate titrations. Acid-base titrations are also required.

Obviously the establishment of a volumetric laboratory does not require big investment. Some glass ware, a few reagents, indicators, distilled water, and reasonably stable temperature conditions are the essentials which can be acquired through a minimum expenditure of about one thousand dollars.

As in the case of gravimetry, numerous methods for volumetric determination of most metals and non-metals have been developed. These can often be used as independent accuracy checks of the gravimetric results. Duplication of a determination by two different methods permits the detection of bias or methodical error, which may be otherwise very difficult to detect.

Titration and colorimetric methods gain special significance in systematic *rapid analysis systems*, such for example as that developed for silicate, carbonate, and phosphate rocks by SHAPIRO and BRANNOCK (1962) at the U.S. Geological Survey. Such systems are used when a large number of samples and a lack of time do not permit gravimetric approach. An establishment of "fast" methods is seldom warranted when only a few samples are to be analyzed because the preparations may require more time than a classical silicate analysis would take.

Volumetric methods have been developed with sensitive indicators which expand the range of sensitivity to cover elements present in minor amounts which otherwise could not be determined directly by gravimetry. Where trace amounts of elements are determined volumetry overlaps today with colorimetry. Limits of detection by volumetric methods vary but we may mention a practical lower limit of 0.1% of the element in a solid sample.

COLORIMETRIC METHODS

In colorimetry we compare the intensity of a color in a sample solution of a known volume with the intensity of this color in solutions of known composition. The absorbance of the solutions is measured by a spectrophotometer, or estimated visually in a colorimeter by bracketing the unknown solution between solutions of known concentration. In this method the intensity of colors of equal liquid columns is compared instead of known volumes being mixed to achieve a certain color change as in volumetry. In photometric titrations the optical density changes of a solution are measured during titration to detect the endpoint.

One selects colorimetric methods for the determination of minor and trace constituents, and in some cases to determine major constituents (Al, Si, and P), especially in industry. These methods overlap mostly into the trace element field but are mentioned here because of their importance in the determination of titanium and manganese in rocks.

FLAME-PHOTOMETRIC METHODS

Flame excitation for quantitative analysis has been recently developed to encompass most of the metallic as well as non-metallic elements. We are interested here in the specific advantages of this method in comparison with other methods used for determination of major elements. In this field we can easily restrict the applications to the alkalies sodium and potassium and the alkaline earths calcium, magnesium, strontium, and barium. The reasons which make the flame-photometric determination of sodium, potassium, and calcium attractive are listed below.

From the text on gravimetric determination of sodium and potassium we see that a great deal of personal skill, effort, and experience is required to perform the multi-stage operations before the alkalies can be quantitatively separated as chlorides or sulfates. On the other hand, to excite sodium and magnesium efficiently by the nondestructive technique a special rather expensive X-ray tube is needed. Every industrial or silicate chemist also knows that the determination of K and Na is one of the basic requirements in most silicate materials analyzed, which means that the analyst may be faced with hundreds of such determinations at a time. In addition, a great majority of smaller industrial laboratories can afford or already possess a simple flame photometer of the Littrow-type (Beckman) or the Perkin-Elmer model-18 series, or similar. (Note that the increased sensitivity achieved by photomultiplier tubes in the more expensive models is not always necessary for the work on major constituents.)

All these facts indicate that in the analysis of Na, K, and Ca (also Rb, Cs, Sr, Ba, and Mg) one will usually find it most con-

venient to resort to the well established and proven flame-photometric or atomic-absorption (pp. 182, 194) methods. Only in laboratories with the necessary X-ray equipment and special tubes (the Al-
target and the Cr-target tubes) already in operation one may switch
over to the nondestructive techniques, thus saving time and the
inconvenience of the HF-decomposition and dilution problems. It is
clear that even in the case of a fully nondestructive analysis setup,
double checks with flame photometry, especially for the elements
mentioned, remain an important necessity in order to be able to
detect a possible bias in the X-ray methods.

With practical detection limits approaching parts per billion in
the trace elements range for the alkalies, the main problem in the
flame-photometric determination of major elements is caused by
errors due to the large dilution factors, and also due to self-absorption and to matrix effects.

ATOMIC ABSORPTION

The main advantages of atomic absorption are high sensitivity
and selectivity, practical absence of spectral interference, simplicity,
and reasonable cost of equipment.

The main disadvantage is the multitude of necessary light sources
which increases the cost. The sensitivity of the method also requires
high dilution factors increasing the possibility of error. The name
"atomic absorption" is often misconstrued to mean a nondestructive method.

When one studies literature on instrumental or classical methods
in the determination of magnesium, it is easy to detect that the
methods for the determination of this element in the presence of
large amounts of calcium leave much to be desired. Only the classical tedious magnesium ammonium phosphate method and the
EDTA titration method, when properly applied, seem to give accurate results. Recently good results have been also obtained in
magnesium determination by Cr- or Al-X-ray excitation, which may
have specific significance for the cement industry.

Atomic absorption seems to give, however, excellent results in relatively wide composition ranges for magnesium. Therefore, if one does not have the expensive vacuum X-ray apparatus for ultra-soft X-ray work, or if one wishes to use a good reference method, it seems that the best solution would be the relatively inexpensive atomic-absorption technique specifically suitable for a fast magnesium determination that is virtually interference-free. Details of this determination are given under the chapter on atomic absorption (Chapter 10).

EMISSION SPECTROSCOPY

Optical emission spectroscopy by electric arc or spark excitation is mainly a tool suitable for the determination of composition ranges from 0.3% to 1 p.p.m. based on the solid sample, and using a photographic plate or direct metered readout. This restriction of the upper determination limit is mainly due to the fact that the gradual darkening of the plate with increased intensity of radiation can proceed only to a certain stage, also to reversal or self-absorption effects with higher concentration. Recently direct reading photomultiplier tubes sensitive in certain specific wavelength ranges have been developed. This has extended the range of this method in many cases to cover higher concentrations of elements as well. Nevertheless, it appears that the main function of this technique will continue to be the detection of elements to be analyzed by other methods. Even though running the risk of offending enthusiasts of this method, one may state that emission spectroscopy is essentially suitable for trace element determination, and as a scanning tool. When applied to major elements it greatly suffers in precision in comparison with the X-ray methods, and, therefore, it here shall not be discussed further.

A comparison of precision of X-ray methods for trace elements in the range of 1,000–10 p.p.m., with precision of emission spectroscopic methods in the same range, shows higher precision to be obtained by the nondestructive X-ray methods, but it should be understood that in this case one seldom needs very high precision. This can be demonstrated by comparing the significance of a 10% relative standard deviation in two substances containing 50% and 100 p.p.m. of the same element respectively. In the first case, working with a 66%-confidence limit, or $\pm 1\sigma$, we could write $50.0 \pm 5.0\%$, which would mean a range of 10% and would not be acceptable unless just a rough estimate was required. In the second case, a total range of 20 p.p.m. would be considered by most spectrographers to be fully satisfactory.

In no way should these remarks discredit the efforts to expand the range of the methods by scientists working in the field of optical spectrography, nor should it be understood as discrediting in any way, this excellent and well established analytical technique. It must be emphasized that equipment for quantitative spectrographic analysis is costly.

CHAPTER 5

Some Methods of Extraction and Ion Exchange Separation

SEPARATION BY EXTRACTION

Extraction of elements from their ionic complexed or chelated solutions with special organic reagents is an old method, and the importance of solvent extraction methods as a means of general separation in complex mixtures is now being fully realized. Analytically the liquid–liquid extraction which uses two immiscible liquids, usually water and an organic solvent, is most used. The solute is transferred from one solvent to the other by shaking for a few minutes in a separator funnel or using more sophisticated equipment. This is called *batch extraction*. The simplicity of this separation is obvious, but one has to control the acidity of the solvent and whenever the concentration of the compound is small, repeated extraction or continuous extraction methods are advisable.

The main organic solvents suitable for extraction are ethyl ether, chloroform, carbon tetrachloride, ethanol, benzene, acetylacetone and their mixtures. By adjusting the pH of the aqueous phase most of the elements can be selectively extracted with these organic solutions. The methods are too numerous to attempt a selection, and the reader is advised to consult the literature list.

It is strongly suggested that the beginner should use extraction–separation methods after acquiring preliminary skill in classical separations as given below. Switching to extraction will often save considerable time, increase the accuracy, and simplify the classical methods by disposing of long evaporation treatments and difficult washing problems.

When using the separatory funnel it is practical to withdraw the liquids into three graduated cylinders, collecting both liquids near the phase border into one small cylinder. This permits the removal of the often impure interphase zone after proper volume estimation in all three cylinders.

In the analysis of common rocks and cements one of the most useful extraction procedures is the *separation of iron by extraction with ether*. This method permits the removal of iron prior to the systematic analysis of other major constituents. The main advantage of this method is that iron and aluminum can be determined directly gravimetrically as oxides, and if other trace co-precipitants are ignored, the determination of aluminum by ammonia precipitation is rendered a direct one, rather than one by difference as is the classical procedure.

Method. Decompose a sample of 100–1,000 mg with HF + $HClO_4$ as given on p. 200, and after evaporating until the disappearance of white perchlorate fumes, treat with 10–20 ml of 6 N HCl, add two drops of 30% H_2O_2 and boil for 1 min. Shake 10–20 ml of the 6 N HCl sample solution with 20–30 ml of ethyl ether in a small separatory funnel for a few minutes. Transfer the acid-water layer with some ether into another separatory funnel, add 20 ml of ether and repeat the procedure twice, collecting the ether (70–90 ml) in another separatory funnel. Strip the $FeCl_3$ from the ether solution by shaking with 10 ml of H_2O twice and once with 5 ml. Boil this aqueous solution under a hood to expel ether and determine iron by precipitation as hydroxide and ignition as Fe_2O_3. Boil the 6 N HCl sample solution to expel ether and use the classical, colorimetric, or ion exchange methods to determine the other main constituents.

SEPARATION METHODS BY ION EXCHANGE RESINS

As in the case of solvent extraction, the development of ion exchange techniques has been revolutionary and so rapid that some inertia still exists among analytical chemists, causing undue delays in adoption of these methods as routine procedures. Consecutive separation of elements by ion exchange is also conventionally referred to as a chromatographic method.

Resins can be used for removal of interfering ions in gravimetric analysis and also for specific sequential separations of elements by the use of eluent solutions of varying composition. A great number of specific, commercially available resins, has been developed. They are divided into two groups; the *cation* and the *anion exchangers* or "absorbers".

The cation exchangers are organic compounds of high molecular weight. They contain sulfonated *acidic* ($-SO_3H$), carboxyl ($-COOH$), or phenolic ($-OH$) groups attached to resins or hydrocarbons. The anion exchangers are also resins or zeolite-like inorganic substances with *basic* amine ($-NH_2$), or quaternary ammonium ($=N-OH$) groups. The active groups of cation exchangers react with ions by exchanging their hydrogens for cations. In anionic exchangers the basic OH^- groups are replaced by anions. If we call the ion exchange resin R, examples of both reactions can be written:

$$R \mid H + Na^+ \rightleftarrows R \mid Na + H^+$$
$$R \mid OH + Cl^+ \rightleftarrows R \mid Cl + OH^-$$

The process of ion exchange is called *sorption*. The bonds between the resin and the ions tend to become stronger when the valence of the ions increases, so that the three-basic phosphoric acid or the three-valent ferric ion (unless complexed) are held more strongly than the respective chlorine or the ferrous iron. This means that separation of different ions is possible by consecutive elutions with acids of different normality. Used resins may be regenerated by passing appropriate acids or bases slowly through the column and washing with water till a neutral reaction to the wash solution is obtained. One to three normal, 1–3 N, hydrochloric acid and 1 N sodium hydroxide are usually used for this purpose. The flow rate usually prescribed amounts to "dropwise" in practical terms. An elution or regeneration of a 60 cm long column with a radius of 1.5 cm may take 30–60 min involving a total of 100–300 ml of eluent. Naturally the best conditions have to be determined by the analyst according to the case. Some easily complexed ions, like Cr^{3+} for example, are strongly bound to the resin, and must be avoided or eluted with strong acids. Hydrochloric acid is naturally not a suitable eluent for silver, and nitric acid must be substituted in this case.

Procedure to determine the capacity of a resin

An ion exchange column is shown in Fig.8. It is usually 10–60 cm long and has a diameter of 1–3 cm depending on the total amounts of cations or anions to be separated. Before filling the column, keep the resin overnight in distilled water. Treat the expanded resin for about 30 min with an excess of either 1–3 N hydrochloric acid if it is an anion exchange resin, or with 1–3 N sodium hydroxide for cation resins. Wash and convert back to either the hydroxyl or the hydrogen form by exchanging the 1 N liquids. Wash a few times with H_2O by decantation and fill the column without packing tightly. Wash through the column until reaction is neutral. If the resin appears irregularly packed, loosen it by pushing water in reverse

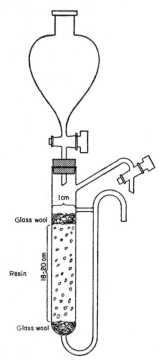

Fig.8. Ion exchange column as used by Professor Hidekata Shibata, Tokio University of Education

direction. Measure the volume of the column. Run a large excess of standardized 1 N hydrochloric acid (anion resin) or 1 N sodium hydroxide (cation resin) solution through the column and back titrate using methyl orange as indicator. The difference in volumes in ml of the 1 N titrants divided by the volume of the expanded resin corresponds to milli-equivalents of the ion per ml of the resin. Usually the ion exchange capacity of modern expanded resins expressed in mequiv. per ml is 1–2 mequiv./ml; it is also given in mequiv./g of dry resin. A salt exchange capacity of a resin is measured by running corresponding salts through regenerated resins.

TABLE II

ION EXCHANGE RESINS IN GENERAL USE

Name	Composition or type	Company
Dowex 50	sulfonated hydrocarbon cation exchanger	Dow Chemical
Dowex 1	strongly basic anion exchanger	Dow Chemical
Dowex 2	strongly basic anion exchanger	Dow Chemical
Dowex-21K	weakly basic anion exchanger	Dow Chemical
Dowex-A-1	chelating exchanger	Dow Chemical
Duolite C-20	sulfonated hydrocarbon cation exchanger	Chem. Process Co.
Duolite A-101	strongly basic anion exchanger	Chem. Process Co.
Duolite A-102D	strongly basic anion exchanger	Chem. Process Co.
Duolite A-101D	weakly basic anion exchanger	Chem. Process Co.
Amberlite, IR-120	sulfonated hydrocarbon cation exchanger	Rohm and Haas
Amberlite, IRA-400	strongly basic anion exchanger	Rohm and Haas
Amberlite, IRA-410	strongly basic anion exchanger	Rohm and Haas

When about 50% of the capacity of a resin is used up, which can usually be observed from a slight change in color, it is advisable to regenerate. Otherwise, unwanted ions may pass into the effluent and the flow rate may be too fast. The resin column always remains under water. Hundreds of specific purpose ion exchange resins are available from numerous sources. Only some common ion exchange resins can be listed in Table II.

In ion exchange work, the length of the column and the volume of the resin, as well as the size of the sample and the volume of the eluent passed through the column, are all interdependent. Whenever the size of the sample or the composition of the sample are varied, one must ascertain that proper separation of ions results, or else change the conditions corresponding to tests for possibly retained ions after the elution with originally prescribed volumes.

For example, it may be necessary to change the respective volumes for the sequential elution of sodium and potassium in the method given below. A rock may contain mostly sodium and little potassium, or vice versa, even though the total alkali content may remain the same. This makes intermittent flame tests necessary. Should one want to use larger than 100-mg rock samples, the whole method given below must be adapted for larger quantities and tested for proper eluent volumes to be used in order to assure quantitative separation of ions.

ION EXCHANGE SEPARATION METHOD FOR SYSTEMATIC ANALYSIS OF SILICATE ROCKS BY SHIBATA

The method described below has been developed at the Geological and Mineralogical Institute of the Tokio University of Education. Professor Shibata has kindly demonstrated this procedure in detail to the author. It has been successfully applied to hundreds of rock analyses. It is straightforward, simple, fast, and reliable, and major elements are separated in the sequence:

$$Fe^{3+}, (P)—Na—K—Ti—Mg, Mn, (Fe^{2+})—Ca—Al$$

Silica is removed with HF during decomposition, and is thus not determined. The separated elements can be determined by either atomic absorption as described (p.194), or combining emission flame photometry, colorimetry, EDTA titration for Ca and Mg (p.88), and oxine precipitation for Al (p.77).

Sample preparation. Add 5 ml of H_2O to 100 mg of fine sample powder, add 3 ml[1] of $HClO_4$, add 5 ml of HF and evaporate on

---

[1] Shibata uses 2 ml of 1/1 H_2SO_4 which is safer than $HClO_4$.

sand bath to near-dryness. Put the platinum crucible or dish aside, cover, add 5 ml of HF and 2 ml of $HClO_4$, evaporate again until all white fumes are gone. Dissolve the residue in 20 ml of 1/1 HCl. Transfer into a dish and boil with 2 ml of 30% H_2O_2 for 5 min to oxidize iron. Evaporate to near-dryness, add 10 ml of 6 N HCl, let dissolve, then cool.

Separation by ion exchange. Prepare two columns (Fig.8), one with Dowex 1–X8 anion exchange resin (100–200 mesh), the other with Amberlite IR–120 cation exchange resin (100–150 mesh) according to general procedure (p.56). Pack about 15 ml of the wet reconstituted resins in each column.

Fe (total)—Pass the 20 ml of the 6 N HCl sample solution through the Dowex 1–X8 column dropwise at a rate of about 1 ml/min. Pass an additional 100 ml of the 6 N HCl solution through the column. Iron forms a chloride complex in this acid solution and is retained on the resin. Elute iron (III) with 70 ml of a 0.5 N HCl solution, test for iron in the following 5-ml portions of this solution with potassium ferrocyanide solution by removing a few drops of the eluent, until no blue color can be detected. From this 6 N HCl solution iron may be determined by any methods commonly used in the laboratory.

P, Na, K, Ti, Mg, Mn, (Fe^{2+}), Ca, Al—Evaporate the *effluent* of the Dowex 1–X8 column to syrupy consistency with crystals present (~ 1 ml), add 30 ml of H_2O, and pass through the Amberlite IR–120 column at 1 ml/min.

P—Pass 30 ml of H_2O containing 2–3 drops of HNO_3 per 100 ml to prevent hydrolysis, combine the effluents, determine phosphorus from this solution after evaporating to 10–20 ml. If much phosphorus is present, up to 100 ml H_2O (+ 3 drops HNO_3) may be used to elute all of it from the anion exchange column. Up to 1% of P in sample can be treated. It may be better to detect P on a larger separate sample using methods given on p.107.

Na—Pass 160 ml of 0.4 N HCl to elute sodium, check the last drops by flame for possible Na. After sodium can no longer be detected, proceed as under K.

K—Pass an additional 150 ml of the same 0.4 N HCl solution to elute potassium.

Ti—Pass 10 ml of 0.8 N H_2SO_4 solution, test with H_2O_2 for yellow color, if no color, add to the potassium-bearing effluent. If color develops, proceed eluting with additional 70 ml of the 0.8 N H_2SO_4 solution. Pass an additional 10 ml of eluent and test color, no color should develop. Determine Ti colorimetrically.

Mg, Mn(Fe^{2+})—Pass 200 ml of 0.8 N HCl. Effluent contains Mg, Mn, and some Fe^{2+} if the oxidation by H_2O_2 has been incomplete. Split the effluent into two equivalent parts for the determination of Mg and Mn; Mg is determined by EDTA, Mn by colorimetry.

Ca—Pass 10 ml of 0.8 N HCl, test for Mg by Eriochrome-black-T, after complexing with EDTA. Pass additional 110 ml of 0.8 N HCl, and test for Al in the subsequent two 10-ml portions of effluent. If no Al is detected by aluminon or the alizarin-S test, add these fractions to the calcium-effluent.

Al—Pass 40 ml of 4 N HCl to elute Al, test the last drops for Al, and if detected, elute with an additional 10 ml of 4 N HCl.

The above system is pictured in a diagram in Fig.9.

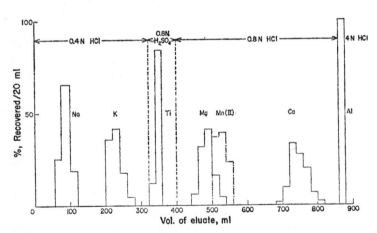

Fig.9. Schematic presentation of chromatographic separation of Na, K, Ti, Mg, Mn, Ca, and Al in analysis of rocks. (Courtesy of Hidekata Shibata, Tokio University of Education.)

CHAPTER 6

The Classical Qualitative Separation Scheme

In the classical silicate analysis the main steps of separation of elements into groups consist of: (*1*) acid treatment to separate silica; (*2*) ammonium hydroxide precipitation in the presence of ammonium chloride to separate iron, and aluminum; (*3*) precipitation of calcium as oxalate; (*4*) precipitation of magnesium as phosphate; (*5*) evaporation to recover alkali metals, which is possible if Mg was first removed as carbonate and if the solution does not contain phosphoric acid. Addition of large quantities of sodium salts during fusion and the presence of large amounts of phosphoric acid after the precipitation of magnesium generally preclude this last step.

As a supplement to this classical quantitative separation, an analyst must remember the *classical qualitative separation scheme* which is very useful in understanding the analytical problems of complex mixtures, especially when the hydrogen sulfide group metals are present. The general outline of a systematic qualitative analysis is given in Table III. It is considered by this author proper to present the qualitative analysis in a book essentially describing quantitative methods, because the systematic treatment of these basic reactions is mostly ignored in modern text books. One seldom finds the word "qualitative" in the indices of most new general textbooks of inorganic chemistry. Thus, it seems that the art of understanding the chemical properties of interacting elements in complex mixtures, the ability to separate, identify, the ability to be on guard for anything unusual, is being slowly lost. The modern chemist is often much out of touch with the basic chemical properties of elements in ionic form, and runs into problems when several elements occur together. Many mistakes can be avoided if the qualitative systematic analytical scheme is remembered.

TABLE III

SYSTEMATIC QUALITATIVE SEPARATION OF ELEMENTS INTO MAIN GROUPS

Elements	Steps of separation
1. Si	Decomposition with HNO_3, or $HF + HClO_4$, evaporation to dryness with HNO_3, insoluble precipitate: Si. Filtrate:
2. Pb, Ba	Precipitation with H_2SO_4, insoluble $PbSO_4$, $BaSO_4$, also some $CaSO_4$, $(BiO)_2SO_4$, Sn, Sb, Sr, Ra. Filtrate:
3. Ag, Hg	Precipitation with HCl, insoluble AgCl, Hg_2Cl_2, also BiOCl, SbOCl. Acid filtrate:
4. Cu, (Pb), As, Sb, Hg, Bi, Sn, Cd	Precipitation in acid solution by H_2S : Cu, Pb, As, Sb, Hg, Bi, Sn, and Cd as sulfides (PbS, if not treated with sulfate). Precipitate of 4 is treated by $(NH_4)_2S$ to separate the CuS and the SnS groups:
5a. Cu, Pb, Hg, Bi, Cd	Residue: CuS, PbS, HgS, Bi_2S_3, and CdS.
5b. As, Sb, Sn	Extract: $(NH_4)_3AsS_3$, $(NH_4)_3SbS_4$, and $(NH_4)_2SnS_3$. Filtrate of 4:
6. Fe, Ni, Co, Zn, Mn, Cr, Al	Remove As if arsenates present by reducing with NH_4I. Precipitate from NH_4OH solution with H_2S: Fe, Ni, Co, Zn, Mn sulfides, Cr and Al hydroxides, also Zr and lanthanide hydroxides. Filtrate:
7. Ca, (Ba), Sr	Boil, remove S by filtering, precipitate with $(NH_4)_2CO_3$: insoluble Ca, (Ba, if not removed as sulfate), Sr, (Mg), carbonates. Filtrate:
8. Na, K, Mg	Contains salts of Li, Na, K, Rb, Cs, and Mg.

Qualitative analysis of a complex mixture starts with acid tests: preferably use nitric acid to decompose the substance, also a mixture of HF and $HClO_4$ if silica is to be removed in rocks. Remember that CaF_2 may precipitate. An insoluble white precipitate in concentrated nitric acid after Na_2CO_3 fusion indicates the presence of Si. Filter, add to the filtrate a drop of H_2SO_4 and stir. If heavy white precipitate forms ($BaSO_4$, $PbSO_4$), add more sulfuric acid drop by drop until further addition does not result in more precipitation and boil. Filter through finest porosity filter. $(BiO)_2SO_4$ also may be present. Remember that Ca may partially precipitate here, also Sr and Ra. Add to the filtrate a drop of HCl and stir. White, heavy, coagulating precipitate which changes color to bluish and brownishblack in the light, indicates Ag; white heavy precipitate indicates Hg_2Cl_2. Also, BiOCl and SbOCl may precipitate. Add hydrochloric acid and stir as long as the precipitate keeps forming, boil and filter through fine porosity filter. Add 5 ml of 5 N HCl. The total volume should be about 200 ml.

Saturate the acid filtrate when hot with H_2S by bubbling the gas through in a stoppered Erlenmeyer flask, as described in Chapter 7 under Zink (p.166). Let stand closed for 2–4 h or better overnight. Black precipitate of *sulfides* may contain HgS, CuS, PbS, As_2S_3, Sb_2S_3, Bi_2S_3, SnS_2, SnS, and CdS. Yellow, orange and brown coloration may indicate As, Sb, and Bi. Wash with a H_2S saturated 5% NH_4Cl solution.

Heat the sulfide group precipitate with 25 ml of $(NH_4)_2S$, stirring continuously on a water bath for 15 min. The residue contains the copper group sulfides of CuS, PbS, HgS, Bi_2S_3, and CdS. The extract contains the tin group sulfides of As, Sb, and Sn in the form of sulfosalts, for example $(NH_4)_3AsS_4$.

If no As is present, add to the copper–tin group *filtrate* 2 g of NH_4Cl and make just alkaline with NH_4OH, heat to boiling and saturate again with H_2S, shake and filter immediately, wash without interruption with hot water containing a few drops of $(NH_4)_2S$ and 5% NH_4Cl. In purely qualitative work these washings may be discarded. This iron group sulfide precipitate contains sulfides of Fe, Ni, Co, Zn, Mn, and hydroxides of Al, Cr, Zr, and the rare earths or lanthanides. If As is known to be present, it must be removed by reduction with NH_4I after double evaporation to dryness with HCl, prior to the iron group precipitation.

Acidify the filtrate of the iron group sulfides with HCl and boil to about half of the volume, filter off the precipitated whitish sulfur, heat to incipient boiling and add an excess of $(NH_4)_2CO_3$ with stirring on the water bath for a few minutes. Filter and wash with hot 1%-$(NH_4)_2CO_3$ water. The *precipitate* contains Ca, Ba, Sr. The *filtrate* contains Mg, Na, and K. Thus, the separation into major groups is accomplished and individual separation of elements to be identified starts. In most cases one will find only a few members of a group present, which simplifies the identification greatly. Any trace elements will, of course, precipitate in the corresponding groups.

The simple methods of identification described under the gravimetric procedures are often the best for quick detection of major constituents. The qualitative system given is of value because it helps to predict or anticipate specific contamination in the quantitative analysis as well.

An excellent and very complete qualitative treatment, including all elements of general interest, is given in a now classical book (LUNDELL and HOFFMAN, 1938). Geochemists and chemists analyzing complex substances and interested in possible co-precipitation of trace elements are well advised to consult this text.

CHAPTER 7

Classical Gravimetric and Other Methods

SILICON

The bulk of the common rocks in the earth's hard crust is siliceous. Thus, about 27.6 wt. % of the lithosphere is composed of silicon. Silicate minerals dominate in most igneous, metamorphic and sedimentary rocks. In the solid crust only the carbonates in the form of limestones or dolomites, and the sulfate or the phosphate rocks may form large bodies where silicon is present in insignificant amounts. Analysis for silicon, Si, or silica, SiO_2, is performed in rocks, minerals, refractories, ceramics and their raw materials, ores, glasses, steel and numerous alloys. Silicon is the most important element in inorganic compounds, just as carbon is in the organic chemistry. In silicates it occupies structurally a central position. In the earth's crust only oxygen is more abundant.

Chemical properties. Silicon, element atomic no.14, belongs to the main fourth group (IV) of elements in Mendelejev's periodic system. It stands below carbon and between aluminum and phosphorus. It has a weakly acidic character. Its oxidation state of four is universal for all analytical purposes. Like carbon, silicon can form giant molecule compounds. In the silicate minerals the most important building block is the silicon-oxygen tetrahedron (SiO_4) with oxygen atoms surrounding silicon in the corners in four-coordination. In silicates, aluminum, phosphorus, titanium, zirconium, and beryllium may substitute for silicon. Generally elements of group-IV exhibit metallic as well as nonmetallic properties, but silicon shows only insignificant baseforming tendencies.

The insolubility of the silicic acid in strong acids is used to separate silica, SiO_2, from other oxides. Characteristic is also the tendency of silicon to form volatile silicon tetrafluoride, SiF_4, when attacked by hydrofluoric acid. This reaction is used to expel this element from its compounds. Water soluble silicates are formed when silicates are fused with alkalies. The *detection* and the *quantitative gravimetric analysis* of silicon are based mainly on these properties.

The presence of flocculent, gelatinous, acid-insoluble silicic acid in the solution after the decomposition of the sample by the sodium carbonate fusion and acid treatment is the simplest practical method of qualitative detection. Only when silica is present in trace amounts the *ammonium molybdate* and *benzidine* test may be required. An analyst learns quickly to detect even milligram quantities of silica during the general process of quantitative analysis.

Gravimetric determination as silica, SiO_2

This determination is applicable, after decomposition, to most ceramics and ceramic raw materials, porcelain, glasses, glazes, portland cement, siliceous minerals and rocks, slags, limestones, magnesites, and aluminum-, chromium-, magnesium-, and zirconium-bearing materials in general, also to bone ashes.

The determination of silica follows immediately after the decomposition of the sample by the sodium carbonate fusion, for example, which is now preferred by most workers to the alternate hydrofluoric-sulfuric acid attack, mainly because the presence of sulfates causes partial precipitation of calcium, lead, strontium, and barium. The sulfate ion is also more difficult to wash from gelatinous precipitates in later stages of analysis.

Procedure

Transfer the cooled cake of the sodium carbonate fusion (see p.26) to a previously used porcelain or better a platinum dish, carefully observing that no matter is left in the platinum crucible or its lid. To assure this, immerse the crucible and the lid in water in a glass beaker and rub clean with a rubber-coated glass rod ("policeman"). Leach on the water bath with water and some alcohol to prevent manganese from oxidizing and forming chlorine which would attack the platinum, until the cake is reasonably decomposed into a soft mass. Add carefully, to prevent effervescence by escaping carbonic acid, concentrated hydrochloric acid, keeping the solution under a watch glass. Make sure that all carbonates are neutralized and that an acid reaction is achieved. Some workers prefer a more rapid method and attack the cake in the crucible directly with 1/1 HCl under cover in a porcelain dish. (Start here with samples decomposed by boric acid attack.)

To assure removal of borates, if these are present, add about 10–15 ml of methyl alcohol, preferably saturated with HCl by bubbling hydrochloric acid gas through it. This procedure is repeated a couple of times during each dehydration process. When a gelatinous greenish crust starts forming in the beaker, break it up by a thick glass rod until a whitish sugary dry powder forms. Complete dehydration can be detected by the disappearance of the hydrochloric acid smell. Keep the dry precipitate under an infrared lamp on a water bath for 20 min and add 20 ml of warm 5–6 N hydrochloric acid, gently breaking the grains with the glass rod until complete dissolution of salts and the separation of the white granular silica at the bottom. Some analysts prefer to moisten the dry residue with concentrated hydrochloric acid and then add warm water. This may prevent partial hydrolysis of titanium salts, if present, but it is questionable whether this variation would have any specific advantage in the general procedure described. This whole process of dehydration may take up to 8 h.

Decant the acid solution through a medium porosity filter paper and wash the silica precipitate seven times with a thin jet of warm 1% hydrochloric acid, each time disturbing and covering the whole precipitate with the wash solution. Collect the filtrate into the same platinum or porcelain dish and repeat the dehydration, carefully "policing" the walls of the dish to assure complete separation of silica which became attached to the walls during the first dehydration. The complete dehydration of the salts is now even more important than during the first treatment and the use of a fine porosity paper is recommended to insure the retaining of small silica particles on the filter. Washing is performed in the same manner as during the first filtration.

Some analysts prefer the baking of the second salt-silica residue at 110° C for a short time to insure insolubility of the silica. This is a procedure to be performed at a temperature not greater than that given because at higher temperatures silica may react with the cations, forming soluble salts. This author prefers heating the dry residue on a water bath with an infrared lamp for 20 min.

If very accurate results are desired, it is better to use the so-called silicomolybdate or the molybdenum blue procedures as given in several texts for the determination of the remaining silica. These methods make the second dehydration of silica unnecessary but require aliquots of the first filtrate. This somewhat disrupts the classical procedure, making it mandatory to bring the first filtrate to a specific volume and then taking the removed portion into consideration in all remaining determinations.

In the *classical method*, transfer both washed silica precipitates, wrapped in their respective filter papers, to an ignited and weighed platinum crucible. Dry the paper carefully, moisten with a few drops of ammonium nitrate solution, dry, and burn with the lid partially open over a gas flame. Ignite the oxide at 1,200° C in an electric oven for at least 30 min. This temperature insures constant weight. It is essential to start the ignition of the filter paper with small flame, otherwise black discoloration of silica will result and graphite or even silicon carbide may form. The paper should never burst into flames during the preliminary ignition. When the oxide is a pure white powder transfer the platinum crucible into a desiccator, cool, and weigh. If the oxide is still hygroscopic, weight increase will be noticed during the weighing and it is advisable in any case to repeat the ignition in the oven until constant weight is achieved.

Volatilize silica by adding to the oxide a few drops of diluted sulfuric acid and a few milliliters of hydrofluoric acid, or if Ba is present, 1–2 ml of perchloric acid in place of sulfuric. Evaporate on a sand bath a couple of times to white fumes. Sulfuric acid is added to prevent titanium from escaping as tetrafluoride, TiF_4. Perchloric acid is supposed to do the same thing. The residue so obtained should not weigh more than a few milligrams after ignition and contains mostly Ti, Nb, Zr, and Ta oxides, but sometimes also insoluble mineral grains with iron and aluminum oxides. Subtract the weight of this residue from the total silica. This residue is later fused with a few grains of potassium bisulfate and the solution obtained made faintly ammoniacal and the washed precipitate dissolved in HCl and added to the combined filtrates of the silica dehydrations. This fusion can be avoided if one waits and ignites the main R_2O_3 precipitate in the same crucible.

In the classical analysis this does not complete the determination of silica. Some silica always remains in solution, no matter how carefully the second silica dehydration is performed, and is co-precipitated with the R_2O_3 group. When a potassium bisulfate or pyrosulfate fusion of the ignited major oxides is performed, this brings the combined weighed oxides of iron and aluminum into solution (see p.75), and renders the *rest* of the silica insoluble when evaporated with sulfuric acid to fumes. This solution is diluted, filtered, and the rest of the washed silica ignited as described before and the weight added to the main silica. Hydrofluoric acid treatment is also advisable at this stage to insure that all of the residue is silica.

Gravimetric factor for silicon:

$$Si = SiO_2 \times 0.46744$$

Interference and contamination. Fluorine, boron, phosphorus, titanium, barium, zirconium, chromium, tin, uranium, tungsten, lead, antimony, niobium, and tantalum, and even calcium, strontium, and magnesium may interfere with the silicon determination or contaminate the precipitate. Incomplete decomposition of certain minerals may also be the cause of error. Before specific explanation of how these different elements may cause errors in silica determination, one must note that fortunately in most silicate rocks the weight fractions of all the trace elements mentioned, except that of phosphorus, titanium, and zirconium, are insufficient to cause serious interference.

Fluorine, if present in amounts larger than 2%, causes low results for silica due to volatilization of silicon tetrafluoride, SiF_4. It may also cause the precipitation of calcium fluoride, CaF_2. This interference is corrected for when the residue from the main silica is ignited and its weight subtracted. Fluorine may cause serious interference also in the determination of aluminum by forming a soluble complex with the hydroxide, and therefore if present in appreciable amounts should be first removed by fuming with sulfuric or perchloric acids.

Boron co-precipitates with silica and volatilizes during the hydrofluoric acid treatment causing high results for silica. It, therefore, has to first be removed with methyl alcohol.

Phosphorus, if present, will also partly precipitate with silica especially with titanium, zirconium, and thorium. It is volatilized as phosphorus pentoxide at 1,200°C causing high results for silica. If a high loss of weight occurs between 1,100°C and 1,200°C there is reason to suspect considerable phosphorus in the silica precipitate.

Titanium is partly co-precipitated with silica and will volatilize with hydrofluoric acid if sulfuric acid is not added, also causing high results for silica. If perchloric acid is used instead of sulfuric acid, titanium is apparently also retained, but this point needs more study.

Barium and strontium may be present as sulfates, zirconium as phosphate or the mineral zircon, if undecomposed. Uranium may be present due to the resistance and insolubility of its minerals in the fluxes used. Tin, tungsten, antimony, niobium, and tantalum may co-precipitate through hydrolysis of their acids in the strongly acid solution. Calcium may be found as calcium fluoride in the silica if much calcium and fluorine are present, and even magnesium may precipitate partly if much phosphorus is present. In all cases when the above mentioned elements are present in specific combinations or appreciable amounts, special precautions must be taken to avoid interference and contamination, and especially when much barium and calcium are present. This is the case with limestones and dolomites. Lead in altered rocks and carbonate-bearing vein material is relatively common, and, if overlooked, may cause serious errors. When much fluorine or phosphorus or both are present in the silicate material, the methods become too involved to be described within the scope of this book.

ALUMINUM

Aluminum is the third most abundant element and the most common metal in the earth's crust, comprising about 8.1 wt. % thereof. In importance as an industrial metal, aluminum rivals iron. Aluminosilicates and hydrated aluminum silicates are widely distributed in nature and belong to the main raw materials and constituents of ceramics, many cements and most mineral building materials. Rare is an analysis where a determination of aluminum is not called for. Nevertheless, the accurate quantitative analysis of aluminum as a major constituent is considered as one of the most difficult tasks in gravimetric chemistry. The main reason for this is the fact that practically every element that may be present in a complex inorganic mixture will precipitate to some degree or will be carried or absorbed by the voluminous aluminum hydroxide precipitate. Aluminum hydroxide is called "Tonerde" in German, and "Ton" also means clay, which is the most common substance in the soils.

Chemical properties. Aluminum, element atomic no.13, belongs to the third main group (III) of elements in the periodic system. It stands below boron and between magnesium and silicon. It often replaces silicon in alumino-silicate structures. Aluminum is much more metallic than boron. Its ionic properties are strong, but its hydroxide, $Al(OH)_3$, is amphoteric, and its salts with weak acids hydrolyze completely in water. There is only one oxide of aluminum in stoichiometric sense, alumina, Al_2O_3. Aluminum hydroxides, $AlO \cdot OH$ and $Al(OH)_3$, precipitate quantitatively in solutions when the basicity of the solution is between 6.5 and 7.5 pH, but due to their amphoteric character begin to dissolve again if one exceeds these limits.

The presence of aluminum is seldom in doubt when the ammonium hydroxide precipitate is bulky. Simple proof is best obtained by boiling with sodium hydroxide, which leaves iron hydroxide undissolved. Then the sodium hydroxide solution can be neutralized, aluminum reprecipitated by ammonia and tested with alizarin or morin. In most cases one rather tests for such possible co-precipitants as the lanthanides or beryllium, because the presence of aluminum is obvious.

Gravimetric method

Applicability, see under Silicon.

The *classical method* uses ammonium hydroxide to precipitate the so-called ammonium hydroxide group of elements, R_2O_3, to which aluminum belongs. These precipitates are voluminous, and gelatinous when filtered, causing co-precipitation of ions as impurities which chemically do not belong here, such as potassium, calcium, and magnesium. Because the aluminum hydroxide is amphoteric and the aluminates are soluble in bases, an excess of the base will cause partial dissolution of the aluminum in the precipitate. In addition to the major elements aluminum and iron, titanium, manganese, zirconium, residual silica, phosphate, vanadium, chromium, the lanthanides, and beryllium co-precipitate. If fluorine is present, it prevents quantitative precipitation of aluminum by this method and must be removed by fuming with H_2SO_4 or $HClO_4$. If not enough ammonium chloride is added to the solution, or if phosphorus is present, calcium and magnesium will precipitate partially. The same is true if the solution contains carbonates and in the case of calcium, if sulfates are present. In the latter case, barium and strontium will interfere as well. The sulfate ion is also absorbed. The determination of alumina by this classical method is indirect because in the combined oxides alumina is determined by difference, based on the determination of iron and the other constituents and contaminants. This means that errors or omissions in the analysis will be reflected in the reported value for Al_2O_3. It is, therefore, advisable that alumina also be determined directly, as shown below (p.76).

Procedure

Combine all filtrates and solutions from the silica dehydration procedures and boil after addition of a few milliliters of bromine water or nitric acid to insure oxidation of ferrous iron. When manganese is present and its complete precipitation is desired, add bromine water to the ammoniacal solution during the precipitation of aluminum hydroxide.

One can alternatively prevent the precipitation of manganese by addition of a few milliliters of ethanol before the precipitation. If only a few percent of manganese are present in the sample and precipitation is performed three times from at least 400 ml of solution, manganese remains in solution and precipitates later with magnesium, if not previously removed.

Usually the amount of hydrochloric acid to be neutralized is sufficient to form enough ammonium chloride to prevent the co-precipitation of calcium and magnesium. But, it is safer to add about 3 g of ammonium chloride to the solution before the precipitation. It is very important to check the basicity of the solution by methyl red, which is, however, quickly destroyed if much nitric acid has been added, or if bromine water was used. Some workers use, therefore, bromophenyl blue or bromcresol purple as indicators.

To this solution which should be at a temperature near boiling and which has been nearly neutralized with ammonia while still cold, and which should be about 250–400 ml in volume, add some ashless paper pulp and then carefully drop by drop of a diluted ammonia solution, until the color change of the indicator just occurs. Stir the solution with a glass rod during this, and boil afterwards for about 2 min. Manganese may precipitate partly if solution is boiled longer, and if no ethanol has been added. Allow the precipitate to settle, and filter still hot through a porous filter.

To insure that no excess of ammonia existed during the precipitation which would have kept some of the aluminum in solution, it is advisable to boil the filtrate for about 5 min, by which time the ammonia odor should only be faintly detectable. Often a small precipitate of aluminum hydroxide forms during this test indicating incomplete precipitation. As the main precipitate, it also has to be filtered off, washed, and redissolved in hot diluted hydrochloric acid.

Washing of the main, often gelatinous hydroxide precipitate has to be performed skillfully. The paper pulp is added before precipitation to facilitate this process, nevertheless if the washing is delayed, or a crust is permitted to form between washings, the time required can become prohibitive. The warm wash solution has to contain ammonium nitrate or ammonium chloride with a drop of ammonia and methyl red indicator to insure basic reaction, because ammonium nitrate solution alone is acid and will dissolve some of the precipitate. Too much ammonia will also cause partial dissolution of the precipitate during the washing procedure.

Wash four to five times, redissolve the precipitate in warm diluted hydrochloric acid, oxidize with bromine water or nitric acid as before, and, after addition of 5 g of ammonium chloride, reprecipitate taking the same precautions and in the same manner. Collect the combined filtrates, acidify with hydrochloric acid, and evaporate to 150 ml.

Because of the interference and the co-precipitation danger mentioned, it is absolutely necessary to perform at least two precipitations, sometimes three if much Ca and Mg is present, or if extra accuracy is desired. One must, however, remember that in the hands of an unskilled worker, excessive reprecipitation may mean accumulative errors. In any case, it is good practice to boil the combined acidified filtrates to about 200 ml and to add dilute ammonia in order to test for aluminum remaining in solution.

The washing of the final aluminum hydroxide precipitate usually takes about one hour, because one must attempt to wash this precipitate until no chloride reaction and only a bare opalescence can be detected with silver nitrate in the wash solution. This test is first performed on a few drops of wash solution passing through the filter after the seventh washing.

Collect the *main and the rest* of the hydroxide precipitates into a fused-silica or platinum crucible and ignite at 1,100°C after the paper has first been carefully destroyed, as described under silicon. The platinum crucible is not used if the substance contains appreciable lead. When equilibrium is achieved, weigh the combined R_2O_3 oxides, crush with platinum rod, mix with potassium bisulfate, and perform *fusion*:

The heating of the crucible after the addition of the bisulfate must start with care because the escaping water and sulfur trioxide fumes may cause the whole substance to boil over. The amount of the bisulfate to be added depends on the amount of oxides present. Usually it is advantageous to try to fuse with as small quantities of the flux as possible: this prevents boiling over and introduces less impurities. Usually about 5 g of potassium bisulfate are used for the main oxide fusion and only a few grains when the silica residue is brought into solution. Some analysts prefer to use pyrosulfate for this fusion, which causes less spattering.

Dissolve the melt in a beaker in dilute sulfuric acid, evaporate to sulfuric acid fumes, cool, add some cold water, and filter the eventually separated silica. Wash, ignite, and after weighing expel silica by hydrofluoric and sulfuric acids. Treat the remaining traces of insoluble oxides, if any, with potassium bisulfate and combine the solution with the main filtrate. This *residual* silica is added to the main silica in calculation of the total silica.

One word of caution must be said here. During the repeated evaporations when glassware, and especially new porcelain dishes, and glass rods are used, it is always possible unknowingly to introduce silica by evaporating ammoniacal solutions before acidifying them, or applying too much force when breaking up the crusts of salts with the glass rods, or by scratching the surfaces of the dishes or glass beakers when mixing or stirring. From this it is clear that the use of platinum ware is to be preferred whenever possible. This, of course, precludes the use of nitric and hydrochloric acid mixtures.

The *filtrate* contains aluminum, iron, and possibly some of the other constituents mentioned in the introduction to this chapter. From this filtrate it is customary to determine total iron as Fe_2O_3, and titanium as TiO_2, and it is good to test for manganese, lanthanides, chromium, phosphorus, zirconium, vanadium, and beryllium, if there is any reason to suspect any of these elements of being present.

This writer prefers to separate manganese by adding a few milliliters of ethanol before the ammonium hydroxide precipitations, which keeps the manganese reduced but does not interfere with the complete precipitation of the iron (III). This permits the separate precipitation of manganese, even if present in substantial quantities, with ammonia and bromine water after the expulsion of all alcohol before the usual oxalate and phosphate precipitation for calcium and magnesium. The manganese is weighed as $MnSO_4$, or better reprecipitated and ignited as $Mn_2P_2O_7$, which permits an extra check on this element. When several percent manganese are present, this separation has to be performed using large volumes of solution.

Alumina is usually determined by difference after the determination of iron from the main filtrate, which is brought for example to 500 ml.

The gravimetric factor to recalculate to aluminum is:

$$Al = Al_2O_3 \times 0.52923$$

A *direct determination* of alumina from an aliquot of this solution can be performed in most cases by the sodium hydroxide method which separates aluminum, chromium, beryllium, phosphorus, vanadium and uranium from iron, titanium, lanthanides, zirconium, nickel, and cobalt.

Procedure for the separation of aluminum in the R_2O_3 group

Pour the faintly acid hot solution containing Al, Fe, etc., into a teflon beaker containing a hot, freshly prepared, 10% solution of sodium hydroxide containing some sodium peroxide (both checked for purity) in such a manner that the volume of the solution is approximately doubled. Stir and boil the solution for about 5 min, permit the precipitate to settle, filter, and wash with a 5% sodium hydroxide solution containing some sodium peroxide, using an alkali-resistant (Whatman 41) filter paper of medium porosity (which has been washed with the same solution discarding the first passed portion). The precipitate can be redissolved in diluted nitric acid for accurate work, and reprecipitated in the same manner. Filter, wash, discard the precipitate, combine filtrates, acidify with hydrochloric acid, and precipitate aluminum hydroxide twice, adding about 5 grams of ammonium chloride as is described above. This procedure leaves aluminum associated with phosphorus and possibly traces of vanadium, uranium, and beryllium. Chromium will also be present if sodium peroxide is used. Wash the hydroxide precipitate, ignite, weigh, and upon checking and correcting for the impurities, especially phosphorus in common materials, report as Al_2O_3.

The total iron can also be determined by titration from the combined filtrates of this separation if retained. This permits a check of the iron determination which is, however, preferably performed directly on an aliquot of the combined R_2O_3 group, taking especially the interference of titanium into account.

If the difference between the sum of aluminum and iron oxides and the total R_2O_3 group is great, there is reason to expect the presence of considerable phsophorus, zirconium, lanthanides, chromium or other elements, and careful qualitative examination is mandatory. In general, spectral examination of the R_2O_3 group is always advisable, and one must remember that no fully satisfactory and simple gravimetric methods for the separation and determination of aluminum are known (see p.54), and that in many cases one has to accept the value by difference when all other constituents in the R_2O_3 group are carefully determined. The importance of the sodium hydroxide separation has, however, in the opinion of the author, been insufficiently emphasized. Even if used only as a check for significant impurities, it gives a simple separation procedure for aluminum in its group. Results obtained by this author when using this method have been satisfactory.

Determination of aluminum by the oxine method

When aluminum has been separated from iron by an ion exchange method (p.54) using a relatively small sample, it may be of advantage to precipitate with 8-hydroxyquinoline instead of using the classical ammonia method. The oxine method is best applicable to samples containing 0.5–50 mg Al. If any of the R_2O_3 group elements are present they will co-precipitate.

Procedure

Neutralize the sample solution with ammonia until aluminum hydroxide just starts precipitating, stir, dissolve the precipitate with a few drops of 1/1 HCl, and add 3–5 drops in excess. The volume should be 50–100 ml. Strongly acid solutions are better boiled to near dryness first to avoid adding large quantities of NH_4OH. Add 5% oxine solution in 10% acetic acid and heat to 80°C. Add 30% ammonium acetate solution drop by drop until a precipitate starts forming, then add 10–15 ml in excess. Cool, check that the pH is between 5.5 and 6.0, if necessary add a few drops of ammonia, let stand for at least 2 h, or better overnight.

Filter through a medium porosity sintered glass crucible and wash with cold water five to seven times. Dry at 130–135°C to constant weight.

Gravimetric factors:

$$Al_2O_3 = Al(C_9H_6NO)_3 \times 0.11096$$
$$Al = Al(C_9H_6NO)_3 \times 0.05872$$

IRON

Iron is the most common metal in the cosmos, and the second most abundant metal in the earth's crust. Considerable iron is present in most substances. This element is also the most important metal in the industry of man. The analytical methods for iron are numerous but only the simplest methods will be described here. Iron is the only major element in nature that occurs simultaneously in multivalent state. Frequently, equivalent values for both of its oxides are desired in order to describe a composition and to perform a proper calculation of totals in a complete analysis. This factor greatly complicates the analytical determination procedures for iron.

Chemical properties. Iron, element atomic no.26, stands with manganese in the middle of the first transition series belonging to the eighth sub-group (VIII) of elements in the periodic system. Members of the so-called iron family, titanium, vanadium, chromium, manganese, iron, cobalt, and nickel, are characterized by the incorporation of new electrons in an inner shell of these atoms, as in the lanthanides and the transuranium elements, which are their heavier congeners. A great similarity in the chemical properties of these elements is so explained.

The highest oxidation state known is Fe^{6+}, it occurs in so-called ferrates, but is rare and of no analytical importance. The main oxidation states of iron are Fe^{2+} and Fe^{3+}. The three-valent state is not as prevalent or stable as in chromium.

The gravimetric analytical chemistry of iron is based on hydrolysis of its salts which starts in aqueous solutions at relatively low pH values of about 3 and the formation of base-insoluble hydrous iron oxides, called often "iron hydroxide" in analytical literature. In fact these are $Fe_2O_3 \cdot nH_2O$, $FeO(OH)$, and similar compounds which on ignition in oxidizing atmosphere lose their water to become Fe_2O_3, which is stoichiometric after ignition at 1,100°C, and is weighed as such. Unfortunately because of co-precipitation of aluminum, iron can rarely be determined in this fashion, and it is best to use the volumetric method of oxidation.

There is seldom doubt about the presence of iron in complex substances. The question of analytical interest concerns mostly the approximate quantity of iron in the sample, which determines the method to be employed. If a *qualitative* test is desired, ferrocyanide or potassium thiocyanate spot tests can be used.

Ferric or total iron, volumetric procedure

Applicable as listed under Silicon.

The acid filtrate of the *residual silica* determination consisting of about 200 ml, or, if separation of aluminum by sodium hydroxide has been performed by taking an aliquot of the combined filtrates after the fusion of the R_2O_3 group, the remaining measured portion of that solution consisting of about 200 ml, is used.

Expel H_2O_2 by boiling. Add about 4 g of ammonium chloride and pass the acid solution through a silver reductor column (p.180), washing with dilute hydrochloric acid, perform the volumetric titration of Fe^{2+} by a potassium dichromate solution (p.174), using sodium diphenylamine sulfonate as indicator.

Titanium and chromium do not interfere as would be the case if the *Jones reductor* (p.180) with amalgamated zinc were used. Copper interferes if present and must be removed by the hydrogen sulfide method. Dichromate titration (p.174) is chosen because of the presence of chlorides after the silver reduction, which would interfere with permanganate titration. If no or just traces of titanium or chromium are present, the use of the Jones reductor and subsequent titration by permanganate in sulfuric acid solution is still to be recommended. Several elements interfere here but in the analysis of materials listed these are present only as traces. Any coloration of the solution passed through the reductors indicates interferences and the cause thereof must be investigated. If little iron is present, it should be determined colorimetrically.

Ferrous iron, volumetric procedure

Applicable to all compounds which are quickly decomposed by a mixture of hydrofluoric and sulfuric acids, and do not contain other reducing agents or organic substances.

Use the —100-mesh grain size powder. If not easily decomposed by acid, regrind to approximately —400-mesh size quickly by automatic vibration mill. Weigh about 500 mg of the fine, just ground powder in a platinum crucible, moisten with about 2 ml of 5 N sulfuric acid and mix with about 10–15 ml of hydrofluoric acid, with platinum wire, heat quickly but carefully to boiling under cover, and permit to boil for about 10 min or just long enough on a sand bath until the material is disintegrated, then immediately wash the contents of the crucible with cold diluted sulfuric acid into a glass beaker containing about 150 ml of a cold solution composed of 105 ml water, 10 ml concentrated sulfuric, 5 ml concentrated phosphoric and about 30 ml saturated boric acid. Rubber gloves should be worn during this procedure. Perform the titration immediately with permanganate solution of a normality close to 0.05 N, or a potassium dichromate solution with indicator.

If longer than 10 min is required for decomposition, the same procedure can be performed under a funnel turned upside down on the sand or water bath, into which CO_2 gas is led to exclude the air. In a dry climate this is not as important as grinding the sample to a very fine powder in a mechanical vibration mill with ceramic sleeves just before the determination, and attacking this powder only for a few minutes until disintegration. All the solutions used should be just boiled to expel atmospheric oxygen, and one must guard against the introduction of organic substances.

The ferrous iron is reported as FeO and also recalculated as Fe_2O_3, which amount is then subtracted from the total Fe_2O_3.

Gravimetric factors:

$$Fe = Fe_2O_3 \times 0.69944$$
$$Fe = FeO \times 0.77731$$
$$FeO = Fe_2O_3 \times 0.89981$$
$$Fe_2O_3 = FeO \times 1.1113$$

Interferences and difficulties characterize the ferrous iron determination. It can be an example of good precision without accuracy just as it sometimes gives good results because the many errors may cancel mutually.

Decomposition of many minerals, especially the oxides chromite, magnetite, and ilmenite, some sulfides, also tourmaline, garnet, and staurolite is incomplete. Most organic substances may reduce the permanganate solution and can cause considerable errors. Sulfides produce hydrogen sulfide which may reduce three-valent iron. Metallic iron from the grinding equipment reduces ferric iron in the sample as well as introduces iron, making the use of ceramic plates in diminution equipment mandatory (see: dual grinding method, p.16). The "last minute" grinding to a very fine powder helps the quick and nearly complete disintegration even of minerals listed here. Naturally, it is only possible in laboratories with vibration type mills with which diminution to an impalpable powder can be achieved in 3 to 5 min.

CALCIUM

Calcium is the third most abundant metal in the lithosphere. It is widely distributed in the rocks, minerals, and waters of this planet, and the living organisms. Limestones, dolomites, gypsum, and apatite rocks belong to the major raw materials of modern industry. In silicate rocks chemically complex calcium minerals dominate. An analysis of a ceramic product, of a rock, of portland cement, or of any ashed organic substance usually requires a calcium determination.

Chemical properties. Calcium, element atomic no.20, belongs to the second main group (II) of elements in the periodic system. Within this group which includes beryllium, magnesium, calcium, strontium, barium, and radium, calcium is the first and most abundant member of the so-called "alkaline-earth" metals, Ca, Sr, Ba. These elements are bivalent in their compounds. Magnesium is also sometimes referred to as an "alkaline earth". This expression originates from the strongly alkaline character of these metals resembling the group-I alkalies, and the similarity of their water-insoluble hydroxide compounds with the hydroxide precipitates of aluminum.

In the "first long period", calcium stands between potassium and scandium. Its normal oxidation state is Ca^{2+}. The alkaline earth metals form a very coherent group chemically. They are all strongly electropositive and possess a pronounced ionic character. In this group chlorides and nitrates are water-soluble, sulfates, carbonates, and phosphates are generally water-insoluble. The solubility of calcium chloride in ammoniacal ammonium chloride-bearing solution, in the absence of carbonates, is used to separate it from the ammonium hydroxide group, R_2O_3. Consecutively the insolubility of calcium oxalate monohydrate, $CaC_2O_4 \cdot H_2O$, in ammoniacal solutions is used to separate and to determine this element gravimetrically.

Calcium is best detected in its compounds when these are inserted into a gas flame on an acid-moistened platinum wire loop. The brick-red flame is observed preferably through a filter. In wet analysis calcium can be precipitated as carbonate after the separation from the R_2O_3 group. Strontium and barium carbonates also precipitate.

Procedure

(*1*) *Ammoniacal solution.* Evaporate the combined acidified filtrates of the double precipitation of the ammonium hydroxide group, after manganese has been removed, until considerable crystallization of chlorides and syrupy appearance, cool, and add about 30 ml of concentrated nitric acid carefully to the glass-covered beaker and permit to react on the water bath until the destruction of ammonium salts is nearly complete. This can be seen when the volume of total salts considerably decreases upon further evaporation to dryness. Dissolve the crystallized salts in diluted hydrochloric acid, bring the volume up to about 200 ml, neutralize with ammonia, add a few drops of methyl red indicator, and make acid again with about 2 ml of excess concentrated hydrochloric acid. Then add 10–20 ml of saturated ammonium oxalate solution, heat the solution to near boiling and while vigorously stirring, slowly neutralize with diluted ammonia, waiting each time after the addition of the ammonia as soon as the crystalline, heavy precipitate starts forming, and avoiding the scratching of the beaker walls with the glass rod. After the methyl red color changes to yellow and the precipitate settles, add additional ammonium oxalate to observe whether the precipitation is complete. Allow to cool for two hours if much magnesium is present, otherwise let the precipitate stand overnight, and filter through a fine porosity filter. Wash four to six times with cold ammonium oxalate-bearing solution, retain the filtrate, redissolve the precipitate in dilute hydrochloric acid and repeat the precipitation as above. Combine filtrates for subsequent magnesium determination.

Wash the precipitate again, and put wrapped in filter paper into a platinum crucible. Char the paper carefully as mentioned under Silicon and finally ignite at full blast of a Meker burner and put into a dessicator containing preferably P_2O_5, or a freshly reconstituted silica gel, for just as long as it takes for the crucible to cool, and weigh, re-ignite, and weigh again as CaO. To recalculate to calcium:

$$Ca = CaO \times 0.71469$$

If the precipitation has been accomplished by very slow drop by drop addition of 0.1 N ammonia, and the washing by the ammonium oxalate followed twice with 5 ml of water, then by absolute alcohol and preferably ether, then the weighing may be done directly as oxalate. In that case a fine porosity Gooch crucible is used, and after drying for 30 min at exactly 105°C, the precipitate is weighed as calcium oxalate monohydrate $CaC_2O_4 \cdot H_2O$.

Gravimetric factors:

$$Ca \ \ = CaC_2O_4 \cdot H_2O \times 0.27430$$
$$CaO = CaC_2O_4 \cdot H_2O \times 0.38379$$

(2) *Acid solution, magnesium rich samples.* Magnesium when present in large amounts as in magnesite, dolomite, or basaltic rocks, may co-precipitate. This is best avoided by an addition of about 20 ml of a saturated oxalic acid solution to about 150–200 ml of neutralized, or faintly acid combined filtrates obtained after the removal of manganese, or, if as in magnesite, the R_2O_3 group is insignificant, directly to the neutralized solution after the removal of silica. The boiling solution is stirred for 3–5 min and examined for precipitates after the addition of the oxalic acid, and if such are detected, carefully acidified with single drops of 5 N hydrochloric acid and stirred until the precipitates are fully dissolved.

Add to this hot solution 40–60 ml of saturated ammonium oxalate, boil for a couple of minutes, and leave the beaker for about one hour on the water bath. Cool, and filter the precipitate through a fine porosity filter, wash with small quantities of a dilute oxalic acid and ammonium oxalate solution, redissolve and reprecipitate this time in ammoniacal solution as above. Combine the filtrates for the determination of magnesium. Ignite the oxalate and weigh as CaO as described above.

Interferences and discussion. Magnesium, manganese, barium, strontium, silicon, aluminum, chromium, iron, lanthanides and platinum interfere by co-precipitating as oxalates, as impurities, or in the form of hydroxides. Sodium and potassium may contaminate the precipitate because, depending on the fluxing methods, one or both of them is usually present in large quantities during the first precipitation. In general, the two methods described above, in combination with proper procedures during the ammonium hydroxide group precipitation, eliminate the possible interference of all the elements mentioned, and the double precipitation in presence of considerable ammonia salts assures the absence of sodium and potassium in the final precipitate.

Because calcium oxalate is appreciably soluble, these methods tend to give slightly low results, especially when washing of the precipitate is overdone.

The first oxalate precipitation may also be performed directly from the combined solutions after the removal of manganese and omitting the destruction of the ammonium salts. In this case reprecipitation under better controlled conditions is mandatory.

MAGNESIUM

Magnesium is the least abundant of the eight main elements of the earth's crust, representing about 2% thereof. It occurs as a major component in ferromagnesian silicates, especially in basaltic and andesitic rocks, and as carbonate in dolomite and magnesite. Large quantities of magnesium are contained in ocean and river waters. In living organisms it is a critical constituent. Industrially it is an important light metal. Analysis of magnesium is normally required in all rocks, ceramic products, refractory materials, cements, clays, carbonate rocks, ocean waters, ashes and a great variety of minerals and alloys.

Chemical properties. Magnesium, element atomic no.12, belongs to the second main group (II) of elements in the periodic system (see Calcium). Magnesium differs in its chemical properties somewhat from the heavier members, as well as from the lighter member of this group, beryllium. The oxidation state is Mg^{2+}. It is highly electropositive but has a tendency to covalent bond formation not present in the heavier homologues of this group. In the absence of ammonium chloride, magnesium hydroxide precipitates in water solutions similarly to beryllium, aluminum, and zinc hydroxides. It is, however, insoluble in the presence of excess ammonia and shows no amphoteric properties as beryllium and aluminum hydroxides do. It is a weak base as compared with calcium, strontium, and barium hydroxides. Magnesium sulfate is water-soluble.

The insolubility of magnesium ammonium phosphate hexahydrate, $MgNH_4PO_4 \cdot 6H_2O$, in ammoniacal solutions is almost universally used to separate it from the alkalies in gravimetric analysis. This is done after the ammonium hydroxide percipitation and the removal of calcium as oxalate, because most of the elements present in an analysis of a complex substance would interfere.

The formation of the crystalline phosphate precipitate after the separation of the interfering ions is also a good qualitative test for magnesium. If zinc is present it must be first removed by hydrogen sulfide precipitation. As with the other major elements, a proof of the presence of magnesium is seldom necessary in the complex mixtures. The main attention has to be directed to the proper identification of impurities, their effect, and removal.

Gravimetric procedure

Acidify the combined filtrates of the calcium oxalate precipitates with nitric acid, evaporate to dryness to destroy oxalic acid. Add about 10 ml of hydrochloric acid, dilute to about 200 ml, and add slowly 20 ml of freshly-prepared solution containing 5 g of ammonium hydrogen phosphate, $(NH_4)_2HPO_4$, while vigorously stirring. Perform the stirring while trying not to scratch or touch the beaker walls by the glass rod, otherwise even at this stage the precipitation of magnesium ammonium phosphate may start. Add concentrated ammonia drop by drop until the solution is ammoniacal, and 5–10 ml in excess. Continue the stirring for a few hours, especially if small amounts of magnesium are present. This is best done by a magnetic stirrer which is then used instead of the glass rod. Let the precipitate stand at least overnight in a cold place, filter through a fine porosity filter, wash with dilute cold ammonia a few times and redissolve in warm, diluted hydrochloric acid using the same beaker with the magnetic stirrer block (these have still some magnesium ammonium phosphate crystals sticking to their sides). See that the combined volume does not exceed 150 ml this time, and add not more than 500 mg of the ammonium hydrogen phosphate, stir to dissolve the reagent, and add drop by drop concentrated ammonia, stirring continuously. When the solution is ammoniacal, add 10 ml of ammonia in excess, stir another hour and let stand preferably overnight, but not less than 4 h. Filter, cleaning the beaker and the magnet carefully with a rubber "policeman", collect all the precipitate on fine porosity filter paper, wash with dilute ammonia six times, dry the precipitate and paper in a platinum crucible, moisten with a drop of ammonium nitrate, destroy the paper very carefully on a small gas flame and ignite. Place finally the magnesium pyrophosphate, $Mg_2P_2O_7$, which now forms, into an electric oven for 30 min at a controlled temperature between 1000° and 1,100°C (but not above). Weigh as $Mg_2P_2O_7$.

Gravimetric factors:

$$MgO = Mg_2P_2O_7 \times 0.36228$$
$$Mg = Mg_2P_2O_7 \times 0.21847$$

Interference and discussion. The magnesium determination concludes the gravimetric classical analysis of the sodium carbonate fusion cake. At this stage practically none of the interfering elements but the alkalies should be present. Calcium may co-precipitate in minor quantities so that in accurate analyses a spectrographic check of the pyrophosphate may be necessary. The main problem is to eliminate the effect of the large quantities of the alkalies present in the solution. The conditions during the precipitation may be such that other phosphates of magnesium, for example $Mg_3(PO_4)_2$ and $Mg(NH_4)_4(PO_4)_2$ co-precipitate. These when ignited do not transform to $Mg_2P_2O_7$, giving stoichiometrically wrong results even though all of the magnesium may have precipitated. The repetition, and the conditions during the second precipitation are, therefore, important to assure that the proper compound, $MgNH_4PO_4 \cdot 6H_2O$, precipitates. If manganese is present it will precipitate as phosphate; strontium will do the same, barium can be tolerated up to at least 5 mg per 100 ml of solution.

EDTA titration methods for calcium and magnesium

Ethylenediaminetetraacetic acid, abbreviated EDTA, and its disodium salt are probably the most useful and versatile organic complexing reagents in modern analytical chemistry. Since the introduction of EDTA by SCHWARZENBACH (1955) and co-workers, some 2,000 papers have been published on the numerous analytical applications of this reagent.

More properly the systematic name of the acid should be spelled ethylenedinitrilotetraacetic acid, according to *Chemical Abstracts*. References to complexon, chelatone, sequestrene, titriplex or versene, and trilon in Russian literature, usually mean EDTA.

The formula of the most frequently used salt, disodium ethylenedinitrilo tetraacetate dihydrate, is:

$$\begin{bmatrix} NaOOC-CH_2 \\ HOOC-CH_2 \end{bmatrix} N-CH_2-CH_2-N \begin{matrix} CH_2-COOH \\ CH_2-COONa \end{matrix} \Bigg] \cdot 2H_2O$$

usually written $Na_2H_2Y \cdot 2H_2O$. This is a white crystalline powder. One hundred eleven grams, 111 g, of this salt dissolve in 1,000 ml of water, whereas the acid form is nearly water-insoluble, 2 g/1,000 ml.

Analytically of prime importance is the formation of complexes of this salt with polyvalent metal ions. This occurs predominantly in a 1/1 molar ratio independent of the valence of the central metal ion. Thus, we can write:

$$Mg^{2+} + H_2Y^{2-} \rightleftarrows (MgY)^{2-} + 2H^+$$
$$Fe^{3+} + H_2Y^{2-} \rightleftarrows (FeY)^- + 2H^+$$

In each case hydrogen ions are liberated, and we can see that the increase of hydrogen ion concentration (or the acidity) of the solution should shift the equilibrium to the left, making the chelate complex less stable. The stability of the EDTA complexes can thus be controlled by adjusting the pH of the solution. Complexometric titrations are based on this property common to EDTA chelates with metal ions. A change of color occurs when excess EDTA reacts with an indicator after the complexing reaction is completed. Ideally this leads to consecutive determination of metal ions from the same solution by the adjustment of the pH only. The calculations of such titration results are simple due to the predominance of the 1/1 molar ratio for most reactions. Thus, we can write:

mg metal sought = (at. wt. metal) × (liters EDTA) × (mol. EDTA)

Generally standardized 0.01 M EDTA solutions are used.

Sample decomposition

Most silicate samples are best decomposed with a mixture of $HClO_4$ and HF. Evaporate to dryness, add 10 ml of $HClO_4$ and evaporate to dryness again. Add hydrochloric acid twice and evaporate to dryness twice to remove fluorine which *masks* calcium and magnesium by precipitation of the fluorides. The residue should be soluble in about 20 ml of warm 1 N hydrochloric acid. Phosphate-bearing solutions are passed through an ion exchange column to remove phosphate ions, which would interfere by precipitating Ca at required high pH values. Most carbonate rocks dissolve by boiling in 5 N hydrochloric acid. Silica residue is then removed by evaporation to dryness and filtration. After the usual treatment with hydrofluoric and perchloric acids the remaining salts in HCl solution are added to the filtrate. Cement can be decomposed by a 5 N HCl solution. *Ammonium salts* should be *absent* from any solution to be titrated first for calcium with EDTA.

Procedure for sequential complexometric EDTA titration of calcium and magnesium using "calcon" as indicator

Neutralize 50–100 ml of the sample solution, which should not contain more than 50 mg Ca, and should contain at least about 10 mg Mg, with 1 N NaOH until a precipitate just starts forming. Add a few drops of a 1% solution of polyvinyl resin, ($—CH_2CHOH—$), in alcohol. Add about 500 mg of KCN, about

200 mg of ascorbic acid, and a few drops of triethanolamine, $N(CH_2CH_2OH)_3$. Mix, and using preferably a pH paper or a calibrated pH meter, adjust the basicity to pH 12.5–13.0 with sodium hydroxide. White magnesium hydroxide precipitates.

Calcium. Heat to about 60°C, add a few drops of "calcon"–sodium-1-(2-hydroxy-1-naphthylazo)-2-naphtol-4-sulfonate—indicator until solution is red and titrate with 0.01 *M* EDTA solution till the endpoint, when the red or pink color changes through purple to blue. Calculate Ca or CaO in mg using the equation given above.

Magnesium. Add to the same solution 3–5 ml of the 10 pH–buffer and boil till the magnesium hydroxide precipitate dissolves. Titrate hot with EDTA after adding a few drops of freshly, prepared 0.1% Eriochrome Black T indicator, sodium-1-(hydroxy-2-naphthylazo)-6-nitro-2-naphtol-4-sulfonate. Wait for permanent color change from red to blue. Add 1 ml of buffer solution and if color changes to red proceed with the titration till the end point. Calculate magnesium from the volume of the 0.01 *M* EDTA solution used, subtracting the added quantity, if any.

Note. When insufficient magnesium is present the endpoint for calcium is not as sharp as desirable for accurate work. It is recommended to add in this case about 5 or 10 ml of magnesium-EDTA solution, prepared by mixing equivalents of 0.1 *M* solutions of these reagents. The detection of endpoints can be improved by photometric means.

If the buffer solution is added before the titration of calcium, both elements can be titrated directly in one step. This procedure is valuable where the total effect or sum of these elements is of importance, as in the determination of *total water hardness*, for example. In that case the addition of the magnesium-EDTA solution is always recommended.

Interference. If major amounts of the sulfide group metals are present, the preliminary removal of these by precipitation as sulfides or extraction in diethyldithiocarbamate is recommended.

Reagents. The 10–pH *buffer solution* is prepared by dissolving about 70 g of NH_4Cl in 450 ml of water and mixing with 550 ml of concentrated ammonia in a polyethylene bottle.

SODIUM AND POTASSIUM

Sodium and potassium are among the eight major elements of the earth's crust. Sodium is more abundant than potassium, comprising about 2.8% of the lithosphere. Sodium and potassium tend to form mainly complex silicates, such as the feldspars for example. These minerals as a group represent about 60% of the bulk of all common minerals in igneous rocks. Feldspars belong also to the main raw materials in the ceramic industry. From the importance of this mineral group, the wide distribution of sodium and potassium in building raw materials can be comprehended. Sodium and potassium oxides are the most critical components in the analysis of ceramics, because their relative amounts influence greatly the physical properties of these compounds. Sodium is also concentrated in the seas and is a vital constituent in all living organisms. As chloride we know it as our table salt. Halite deposits have formed through evaporation of salt lakes and sea water. In most total analyses the determination of sodium and potassium as major constituents is required.

Chemical properties. Sodium, element atomic no.11, and potassium, element atomic no.19, belong to the first major group (I) of elements, which include lithium, sodium, potassium, rubidium, and cesium. They are collectively referred to as "alkali" metals and form a very coherent group chemically. The strongly ionic character of their univalent, positive ions, and the pronounced basic character of their hydroxides are typical. Their atoms have only a single s-electron outside of the inert-gas electron configuration. This determines the simplicity of their chemistry. The polarization tendency of their univalent ions is weak, because they possess the inert-gas configuration and are thus spherical, resulting in chemistry which is based on the univalent positive ions of these elements. This similarity leads to difficulties in analytical separation of these elements. The most important chemical properties of alkali metals are the solubility of their salts with all the acids in general use in a laboratory. Their hydroxides are also water-soluble. In fact there are only a few complex compounds that form insoluble or poorly soluble salts with alkalies. As a result, sodium and potassium remain in solution till the final stage of the silicate analysis, because they stay in the filtrates, and finally are precipitated as chlorides or sulfates through evaporation of the liquid. One uses the relative insolubility of potassium chloroplatinate, K_2PtCl_6, potassium perchlorate, $KClO_4$, or sodium zinc uranyl acetate hexahydrate, $NaZn(UO_2)_3 \cdot (CH_3COO)_9 \cdot 6H_2O$ to separate either potassium or sodium from their mixtures.

Qualitative detection of sodium and potassium is best performed using the coloration their ions give the gas flame. Typical is the easy excitation of their emission spectra at relatively low temperatures. Simplicity of these spectra helps in the flame-photometric determination of these elements. The bright yellow sodium flame, and the purple color of the potassium flame are familiar to all analysts.

Decomposition of sample

Carbonates and other acid-soluble substances like some cements may be dissolved directly in hydrochloric acid. When this is done it is sometimes advisable to add 3–5 g of ammonium chloride after decomposition in order to form enough bulk salts which facilitate the first dehydration of silica, which has to be removed first and is then performed in the classical manner. In this case the third group still has to be precipitated, followed by the removal of Ca and Mg. Since all of these procedures take time it is often better to perform the decomposition of the sample by the *J. Lawrence Smith method* given below, unless insufficient material is available, or if one wishes to perform the complete analysis of major constituents on one sample. A *dissolution* method suitable for alkali determination is an attack by a mixture of hydrofluoric, nitric, and sulfuric acids, or hydrofluoric and perchloric acids:

In a teflon beaker, moisten and mix the impalpable powder with 1–2 ml of 5 N sulfuric acid or with 2 ml of concentrated perchloric acid instead, add 10 ml of concentrated hydrofluoric and, if sulfuric is used, 1 ml of concentrated nitric acid, evaporate to white fumes. If the sample is not fully decomposed, again add 10 ml of

HF and 1 ml of HNO_3 if sulfuric acid is used, and evaporate to dryness. The residue, if not soluble in warm hydrochloric acid has to be examined and identified. The separations after this follow the classical steps, with the difference that magnesium has to be separated first as hydroxide from concentrated solution to avoid the introduction of phosphate. In general this decomposition, especially when performed with perchloric acid, is best suitable for further flame-photometric determinations of the alkalies and the alkaline earths, where the presence of magnesium does no harm.

J. Lawrence Smith method

Procedure (applicable to materials listed under Silicon)

Mix 500 mg of very finely ground sample, preferably 95% of which passes a 400-mesh screen, thoroughly in an agate mortar with 500 mg of ammonium chloride and about 3.5 g (5 g if much iron is present) of pure anhydrous calcium carbonate, $CaCO_3$, especially prepared for this purpose. If the sample has insufficient silica, a better sintered cake is obtained if 1 g of pure, fine quartz powder is added. A finger-shaped platinum crucible, approximately 10 cm long and 1.5 cm wide is preferably used. Cover the bottom of the crucible with about 0.5 g of pure calcium carbonate, add the main mixture, clean the agate mortar and pestle carefully, and "wash" twice with about 0.5 g of pure calcium carbonate. Cover the powder in the crucible with about 1 g of the reagent, $CaCO_3$. Heat the crucible with lid almost closed in an inclined position, first with a small gas flame, increasing the heat gradually until ammonia fumes can not be detected, and then closed, over full blast of a Meker burner for 1.5–2 h. The sintered button is usually grey or brown, and should show no fused glassy appearance at the bottom if enough carbonate and silica were present.

Wipe the cooled crucible from outside and place into a platinum or porcelain dish and let the cake decompose covered with water over a water bath. It is better to decompose overnight. Break up lumps carefully, and filter, after decanting and leaching with small amounts of water about three times. Wash the precipitate with warm water, preferably saturated with calcium hydroxide. If much mica was present in the original sample, collect the insoluble portion, dry, mix with 500 mg of NH_4Cl and resinter. Evaporate the combined filtrates to about 50 ml, add 5 ml of concentrated ammonium hydroxide and about 2 g of solid ammonium carbonate. Boil for a short time, filter through a fine porosity filter under suction into a platinum dish, wash with hot water using about 3 ml-quantities about four times, and evaporate to dryness. Sublimate the ammonia salts carefully, dissolve the remaining chlorides in about 10 ml of water made just acid with hydrochloric acid and add a few drops of barium chloride solution. Precipitate the sulfate ions, and let stand on water bath for 10 min. Add a few drops of saturated oxalic acid, followed by a few drops of ammonium hydroxide, and about 0.5 g of ammonium carbonate. Digest the ammoniacal solution on water for about 20 min, cool, filter through a fine porosity filter, and wash with small quantities of 1% ammonium oxalate solution. Evaporate the filtrate to dryness, add about 10 ml of water, and filter off the insoluble residue. Make the solution acid with a few drops of hydrochloric acid, evaporate in a platinum dish, carefully heat to sublimate the ammonia salts and destroy the carbonates formed from the oxalic acid with a few drops of hydrochloric acid. Finally heat with a gas flame to dull red, moving the gas flame quickly, in order to just barely melt the chlorides all around the crucible walls. Cool in a desiccator, and weigh the combined chlorides. Dissolve again in water, filter off the residue, and repeat the whole procedure, remembering each time to add a drop of hydrochloric acid before the gentle ignition of the chlorides, until the insoluble brownish residue that forms each time is barely detectable on the platinum surface. This may require several repetitions depending on the skill of the analyst.

Sodium and potassium are now in the form of chlorides. Lithium, rubidium, and cesium, also some calcium and magnesium may be present, especially the latter two, if the separations have not been carried out carefully using small volumes of washing solutions. The combined chlorides are weighed at least three times between the last treatments in order to assure a more or less constant weight. Recent thermogravimetric work has shown that both chlorides are stable till the temperature rises above 800°C. Drying at 500°C to remove all occluded water may therefore be recommended.

Depending on agents available, the potassium is precipitated as perchlorate, potassium tetraphenylboron, or as chloroplatinate. Sodium may also be determined as zinc uranyl acetate. Both of these alkalies can now also be determined by flame-photometric methods which permit simultaneous determination of other alkalies, and at least a check for possible impurities.

Potassium perchlorate method

This author considers the method employing the insolubility of potassium perchlorate in absolute alcohol (better in n-butanol) as satisfactory if performed skillfully. It is the simplest and the least expensive of the gravimetric procedures. Generally, however, this method is considered less satisfactory than the chloroplatinate precipitation.

Procedure

Evaporate the combined chlorides with perchloric acid twice in a platinum dish on a water bath to syrupy mass consisting mostly of crystals. Add each time about 3 ml of 70% perchloric acid, then add about 20 ml of absolute alcohol or anhydrous n-butanol (better a 1/1 mixture of n-butanol and ethyl acetate), to which a few drops of perchloric acid have been added. Digest on water bath for 20–30 min while mixing occasionally, decant through a sintered glass crucible. Wash a couple of times with the same solution, redissolve in water and evaporate with a few drops of perchloric acid to fumes, filter through a sintered glass crucible, wash with the same alcohol solution, dry first at 110°C, covered to prevent losses, then increase the temperature of the oven to about 350°C for some 20 min, cool, and weigh as $KClO_4$. Repeat heating at 350°C for 20 min and check the weight.

Gravimetric factors:

$$K \ \ = KClO_4 \times 0.28222$$
$$KCl = KClO_4 \times 0.53810$$
$$K_2O = KClO_4 \times 0.33993$$

Discussion. Lithium follows sodium in this method and may be determined by flame photometry afterwards. Recent thermogravimetric work has shown that anhydrous potassium perchlorate starts to decompose only above 600°C, but that certain impurities may catalyze the decomposition.

Since organic solutions are used here with perchlorate, work should be performed behind shatterproof glass in a perchlorate hood to avoid and guard against possible explosions.

Potassium tetraphenylboron method

It is characteristic of the nature of analytical chemistry that even an excellent and highly selective method, when new, often takes an unduly long time until it gains general acceptance after its initial introduction. The precipitation of potassium tetraphenylboron, $KB(C_6H_5)_4$, which was first described by Wittig in 1949, is a good example of this inertia.

Procedure

The sample should not contain less than 2 mg nor more than 25 mg of potassium. This range covers most rocks if a 500-mg sample is taken. Potassium feldspar and other potassium-rich materials must be taken in proper quantities. For this purpose the solution of the combined chlorides may be split and diluted. The best results are achieved when about 20 mg of potassium is present.

Dissolve the chlorides in about 10 ml of water in the same platinum dish, in which the weighing was performed. Transfer into a divided flask watching that the combined wash solutions do not exceed 45 ml in volume. Add 4 ml of 5 N hydrochloric acid and adjust the volume to 50 ml. Cool the contents of the flask in ice water and add 25 ml of ice-water-cooled 1% sodium tetraphenylboron solution. Continue swirling for a few minutes and after about 15 min filter through a fine porosity sintered glass crucible. Do not let stand for longer than an hour. Wash three times with a fine jet of a saturated cool potassium tetraphenylboron solution, also rinsing the flask, and seeing that not more than about 30 ml of the wash solution is used, preferably in 5-ml portions. Dry the precipitate for an hour at 110°–120°C, cool in a desiccator and weigh.

Gravimetric factors:

$$K \quad = KB(C_6H_5)_4 \times 0.10912$$
$$K_2O = K \times 1.2046$$
$$KCl = K \times 1.9068$$

Interferences. This tetraphenylboron method is remarkably free of interference. Only ammonia, rubidium and cesium co-precipitate. Titanium, univalent copper, thorium, and cerium of the trace elements commonly present in the materials treated here also precipitate, but should not be present at this stage. In addition, mercury, thallium, and silver interfere. Because aluminum, iron, calcium, magnesium, manganese, and chromium do not interfere, this precipitation can also be performed at an earlier stage of the silicate analysis if only potassium has to be determined. The introduction of the sodium salt prevents, of course, further work on sodium.

Reagents. Sodium tetraphenylboron reagent: shake 3.5 g of sodium tetraphenylboron in 100 ml of water in a closed flask; add about 1 g of moist aluminum hydroxide gel and shake mechanically for about 15 min; filter through a medium porosity filter till solution runs clear, switch beakers, refilter this and the remaining portion through the same filter paper. The solution should be clear. This amount of the reagent should be sufficient for four determinations.

Wash solution: dissolve about 50 mg of analytical grade potassium chloride in about 100 ml of water, add about 1 ml of 0.1 N hydrochloric acid and precipitate by adding the sodium tetraphenylboron reagent drop by drop while stirring; filter after 15–30 min, wash several times with small quantities of cold water, dry the precipitate at 120°C, and store in a closed weighing bottle; to make the wash solution, vigorously shake about 30 mg of potassium tetraphenylboron in about 250 ml of water for about two hours, filter, and use when fresh.

Sodium as uranyl zinc acetate

As stated above, sodium is usually determined by difference or by flame photometry, which procedure is to be recommended. If an accurate gravimetric determination of sodium is desired, the combined chlorides can be dissolved in a 100-ml volumetric flask, and separate aliquots containing the desirable amounts of the elements pipetted for the precipitation of potassium as tetraphenylboron salt and sodium as uranyl zinc acetate salt.

Procedure
Method of Barber and Kolthoff. Because of the solubility of the sodium salt, $NaZn(UO_2)_3(C_2H_3O_2)_9 \cdot 6H_2O$, it is important that all solutions are kept at the same temperature.

Evaporate 10–30 ml of the neutral chloride solution, representing not more than 8 mg of sodium to about 1 ml, add 10 ml of zinc uranyl acetate solution, stir for about 1 h mechanically, permit to settle, filter through a fine porosity sintered glass crucible by decanting first and transferring the precipitate to the crucible with as small quantities of the alcohol wash solution as possible, and wash with ether or acetone about three to four times. The dry, clean crucible is put on the balance and weighed in about 10 min, repeating the weighing twice.

Gravimetric factors:

$$Na\ \ = NaZn(UO_2)_3(C_2H_3O_2)_9 \cdot 6H_2O \times 0.014948$$
$$Na_2O = NaZn(UO_2)_3(C_2H_3O_2)_9 \cdot 6H_2O \times 0.020149$$
$$NaCl = NaZn(UO_2)_3(C_2H_3O_2)_9 \cdot 6H_2O \times 0.038001$$

Reagents. To prepare uranyl zinc acetate reagent, make two solutions: (*1*) add 4 ml of glacial acetic acid to 100 ml of water and dissolve about 20 g of uranyl acetate, $UO_2Ac_2 \cdot 2H_2O$, by warming; (2) add 2 ml of glacial acetic acid to another 100 ml of water and dissolve about 60 g of $ZnAc_2 \cdot 2H_2O$. Mix the solutions, add a few drops of dilute sodium chloride solution, stir and let stand overnight. Filter through a dry, fine porosity paper into a dry bottle.

Alcohol sodium zinc uranyl acetate wash solution. Precipitate the sodium salt under conditions given under procedure, wash, shake with absolute alcohol, and filter the next day.

TITANIUM

Titanium is the most abundant minor element in the lithosphere; its average content in igneous rocks is about 0.5%. This metal has important metallurgical uses due to its resistance to corrosion. Titanium oxide, TiO_2, is used as a pigment in paints. In most rock, mineral, and ceramic material, titanium analysis is required. The main titanium minerals are ilmenite, $FeTiO_3$, rutile, TiO_2, sphene or titanite, $CaTiSiO_5$, and titanian magnetite, $Fe(Fe,Ti)_2O_4$. In the case of titanium, one can say that a truly universal method for its determination exists.

Chemical properties. Titanium, element atomic no.22, stands at the beginning of the first transition series between scandium and vanadium. It belongs to the fourth sub-group (IV) of elements in the periodic system, which also includes zirconium, hafnium, and thorium. All elements in this group are dominantly tetravalent and similar in their chemical properties. Some similarity with silicon exists. Titanium can substitute for silicon in many silicates due to similar ionic size and the same valence state. Although titanium can occur in oxidation states I, II, and III, the most stable compounds are tetravalent, IV. Basic oxysalts or hydrated titanium oxides precipitate with ammonia, and titanium is quantitatively retained in a combined iron and aluminum "hydroxide" precipitate. The most characteristic reaction for titanium is the formation of an intense reddish-brown to orange-brown coloration when peroxides of alkalies or hydrogen are added to acidified solutions containing

Ti^{4+} ions. The colored peroxytitanyl compound is destroyed in the presence of fluorine ion. Quantitative chemistry of titanium (when in concentrations of less than 3%) is based on the colorimetric or spectrophotometric determination of the relative intensity of this coloration, or its specific absorbance.

Titanium is easily identified in H_2SO_4 solutions by this reaction in the absence of interfering ions, which are iron, chromium, and nickel in larger quantities because of the color of their solution, and vanadium, molybdenum, and niobium which form similarly colored compounds with peroxide. HCl should be absent because of a tendency to form colored complexes with some metal ions.

Titanium as peroxide

Applicable to all materials listed under Silicon.

Either an aliquot of the dissolved bisulfate fusion of the R_2O_3 group, or the same solution from which iron has been determined, is used. An independent sample can be fused with sodium carbonate or potassium bisulfate, leached with sulfuric acid, silica removed, the residue fused and added to the main portion, and treated with peroxide.

Procedure

Bring the acid solution to approximately 70 ml, transfer it to a volumetric flask of 100 ml, add about 5 ml of concentrated sulfuric acid and about 10 ml of a 3% hydrogen peroxide solution, permit the color to develop for 5 min on water bath, and add another 5 ml of peroxide, observing whether any darkening of color develops. If color darkens considerably, the liquid has to be transferred to a larger flask, adding enough sulfuric acid to keep it at approximately 5%, and enough peroxide. When the color has fully developed, comparison can be made with a standard solution for absorbance, preferably with a spectrophotometer at 410 mμ.

Interferences are many (see above), but in the analysis of materials listed, only the following precautions need be taken.

If iron, chromium, and nickel are present, the approximate absorbancy of their combined interference can be measured before the addition of peroxide to the nearly full flask (remember to return the portion used) at the wavelength to be used during the determination, and the total effect subtracted later. The addition of similar amounts of persulfate to the standard solution also improves the results. If colored compounds of vanadium, niobium, or molybdenum are suspected, these have to be removed by precipitating titanium with iron by sodium hydroxide before the determination. Addition of phosphoric acid to weaken the color of the Fe^{3+} ion may precipitate titanium if it is present in large amounts. Fluorine interferes, and has to be fumed off with sulfuric acid.

Preparation of standard titanium sulfate solution. Use standard grade TiO_2, ignite in fused silica crucible, cool in dessicator, and weigh the desired quantity, usually 1,000 mg. Fuse with 10 g of $KHSO_4$ till the flux is clear, cool, and dissolve in 200 ml of cold, 25% solution of H_2SO_4 in a 1-liter volumetric flask. Dilute to the mark with distilled cold water kept at room temperature. This solution corresponds to 1 g TiO_2/l in an approximately 4–5% H_2SO_4 solution.

MANGANESE

The most common trace elements in rocks are manganese and titanium. Within the so-called group of heavy metals, only iron is more abundant in crustal rocks. Chemically manganese is related to iron, and is usually associated with this element in nature. The determination of manganese is a relatively simple matter due to the strong coloration of the permanganate ion. Because of this, manganese, even though present as a trace or minor element, is determined in most analyses and reported with the major constituents. In fact, most rock analyses must be considered incomplete if manganese is not determined.

Chemical properties. Manganese, element atomic no. 25, stands in the middle of the first transition series between chromium and iron. It belongs to the seventh sub-group (VII) of elements. The highest oxidation state of manganese is Mn^{7+}. The relative ease of reduction of this highest oxidation state, combined with its strong color, are used to *determine* this element in its solutions by colorimetric means, comparing the relative intensity of the color of the permanganic acid. The Mn^{2+} state is the most stable, therefore, complete reduction of a permanganate, $KMnO_4$, for example, is the goal of an analyst.

Analytically, the Mn^{4+} oxidation state is also of importance. The quadrivalent hydrous dark-reddish-brown manganese oxide is precipitated with a small excess of ammonia when strongly oxidizing agents are present.

Manganese is best detected by the typical permanganate coloration produced when its compounds are treated by sodium bismuthate, potassium periodate, or ammonium persulfate in cold nitric acid solutions.

Separation and determination of manganese

Applicable to all materials listed under Silicon.

If two precipitations by ammonia have been performed in the presence of a few milliliters of ethyl alcohol, from solutions about 400 ml each, practically all manganese is in the combined filtrates. It is advisable to separate it before the precipitation of calcium and magnesium (p.76).

Procedures

Evaporate the combined acidified filtrates to about 100 ml.

(*1*) *Low manganese, precipitate to be discarded.* Add 5–10 mg zirconyl chloride in solution, nearly neutralize the solution with ammonia, add persulfate and macerated paper, heat the solution, make ammoniacal, and reheat again. The precipitate consisting of hydrated Mn^{4+} oxide and zirconium hydroxide is filtered off, washed, and usually discarded. It may be redissolved in dilute sulfuric acid, oxidized, and the manganese determined by the colorimetric method, if insufficient material is present. Manganese, however, is usually determined from a separate sample.

(2) *Over 0.5% manganese to be determined gravimetrically*. In this
case, no zirconyl salt is added, the alcohol is boiled off, and the
precipitation performed with ammonia in the presence of bromine
or persulfate, the washed precipitate dissolved in hydrochloric
acid, and $MnNH_4PO_4$ precipitated as described under magnesium,
ignited, and weighed as $Mn_2P_2O_7$.

Gravimetric factors:

$$Mn = Mn_2P_2O_7 \times 0.38713$$
$$MnO = Mn_2P_2O_7 \times 0.49988$$

The main advantage of determining manganese at this point lies
in the fact that practically no other interfering elements are present.

Colorimetric determination of manganese

This method is given here because it is used most commonly in
analyses of materials listed.

Oxidation of manganese to the permanganic acid may be performed by periodate. The *periodate method* is preferred because,
unlike with bismuthate for example, no filtration is required, and
the permanganate in the presence of an excess of periodate is stable.

Procedure

An aliquot of the combined evaporated filtrates, after the silica
determination or after the dissolution of the R_2O_3 group, if all
manganese has been co-precipitated in basic solution by adding
bromine, can be used. No chlorides should be present. Potassium
periodate or sodium periodate is added in excess to an aliquot of a
boiled solution containing about 10% sulfuric, 20% nitric, and about
5% phosphoric acid, boiled for a couple of minutes, and left on a
water bath for 10–15 min to permit full oxidation. The solution is
cooled and diluted to a desired volume. Then the intensity of the
permanganate color is compared with solutions containing similar
amounts of the acids and known amounts of permanganate. This
can be done satisfactorily in a colorimeter of the simplest type. If
chromium is present, as may be the case in some refractories and

minerals, or rocks, the comparison may be done photometrically at a wavelength of 545 mμ.

Reducing agents, separated silica, and organic substances in general interfere with all oxidations to permanganic state.

PHOSPHORUS

Phosphorus belongs with manganese and titanium to the most abundant trace elements in the lithosphere. Like the other two elements mentioned, phosphorus is conventionally determined in rocks, ceramic materials, ores, slags, alloys, steel, bone ashes, and portland cement. In phosphate rocks, such as phosphorite or apatite, phosphorus is present as a major constituent. In living organisms, phosphorus is of vital importance, especially in bones. Plants also need phosphorus. In general, phosphorus does not show any pronounced tendency to concentrate in either the silicate, the metal, or the sulfide phases of the physico-chemical systems characterizing the planet Earth. Therefore, it is widely distributed among the rocks and their minerals. Magnesium- and iron-rich rocks generally carry somewhat more phosphorus than the more siliceous members. In the sedimentary cycle, phosphorus follows calcium mostly due to the insolubility of calcium phosphates, which is complemented by the occurrence of these phosphates in bones of animals and decaying organisms.

Chemical properties. Phosphorus, element atomic no.15, belongs to the fifth main group (V) of the periodic system. This group consists of nitrogen, phosphorus, arsenic, antimony, and bismuth. Phosphorus stands between silicon and sulfur in the third period. Like nitrogen, it is mostly covalent in its compounds, whereas As, Sb, and Bi show increasing metallic character. Its oxidation state of five against oxygen is dominant. Phosphorus pentoxide is acidic, but the metallic character of its homologues increases toward bismuth. In minerals of the lithosphere, phosphorus occurs as orthophosphate, which is its most stable compound. Orthophosphate is analytically the most important ion. Most phosphates are soluble in strong mineral acids. The analyst must, however, expect other acids of phosphorus and their salts. For example, pyrophosphates, metaphosphate glasses, and polyphosphates. In case any of these are suspected, special decomposition techniques must be used. The phosphate anion PO_4^{3-} forms with the molybdate ion an acid-insoluble yellow complex, the so-called ammonium phosphomolybdate of the empirical formula $(NH_4)_3(PO_4 \cdot 12MoO_3) \cdot 12H_2O$. This compound is used to precipitate phosphorus in nitric acid solutions in order to separate it from other ions. Another insoluble compound used in gravimetric determination of phosphorus and magnesium is magnesium ammonium phosphate hexahydrate, $MgNH_4PO_4 \cdot 6H_2O$, which upon ignition is transformed into magnesium pyrophosphate, $Mg_2P_2O_7$, which is a stoichiometric compound.

In the qualitative detection of phosphorus the same reactions and the yellow color of the molybdate compound are used. As with other major elements and major trace elements, there is seldom doubt about the presence of phosphorus in at least trace amounts in a sample submitted for analysis, and again the problem is to identify the interferences and the contaminants. Silica also forms a yellow ammonium silicomolybdate compound.

Gravimetric determination

In classical silicate analysis a separate sample is used.

Solution of the sample. Applicable to rocks, minerals, clays, most ceramics, and carbonate samples.

Ignite 1 g of finely ground sample powder in platinum crucible over a gas burner for about 20 min to destroy organic substances and carbonates, if present (if sulfates are present, do not ignite). Cool, and moisten the powder, cover with platinum lid, add 5 ml nitric and 10 ml hydrofluoric acid, mix, let stand on water bath covered until strong reaction ceases, and evaporate to dryness. Repeat once, add 5 ml of concentrated nitric acid twice, and evaporate to complete dryness each time. Dissolve the contents by adding 20 ml of concentrated nitric acid, digest the precipitate till it separates from the crucible walls, transfer into a teflon beaker by washing clean with dilute nitric acid, add 10 ml of saturated boric acid solution and digest covered with teflon for about 2 h on a shaking sand bath, adding more concentrated nitric acid if necessary to keep the volume to about 50 ml. Filter into a 100-ml volumetric flask through a fine porosity filter and bring the liquid up to the mark when cool.

Phosphorus and manganese can be determined from aliquots of this solution. If the sample contains considerable titanium, zirconium, or thorium, the insoluble residue is mixed with sodium carbonate after the acid treatment and, after fusion, is leached by water, filtered, filtrate acidified with nitric acid and added to the main portion. When tungsten is present, this solution will contain most of it and should then not be added to the main solution.

Samples soluble in acids, as bone ashes, fertilizers, carbonate rocks, most orthophosphates, and phosphorite ores, are dissolved by nitric acid digestion. This method does not leave much fluorine in solution, but it is still advisable, since apatites contain fluorine, to add a few milliliters of boric acid to the solution to assure quantitative precipitation of phosphorus as ammonium phosphomolybdate.

Samples containing condensed phosphates, for example polyphosphates and metaphosphate glasses, have to be hydrolyzed, after dissolution, to orthophosphates by boiling to dryness with hydrochloric acid in the presence of about 1 g of potassium chloride, KCl. The residue is dissolved in nitric acid, evaporated to dryness twice, and redissolved in nitric acid.

Metallurgical products, steel, ferrous and non-ferrous alloys, contain phosphorus as phosphide. If attacked by non-oxidizing acids, phosphine gas would volatilize causing low results. The acid attack has to be performed then by dilute nitric acid with addition of potassium permanganate to a boiling solution until a dark brown precipitate of manganese dioxide forms.

Ammonium phosphomolybdate method

Ammonium phosphomolybdate, precipitated in nitric solution, is generally used to determine phosphorus. The yellow molybdate precipitate can be weighed as such. If more than 0.5% phosphorus is present, the precipitate is redissolved and phosphorus reprecipitated as magnesium ammonium phosphate, or titrated. This method is almost of universal nature and is especially suitable for the analysis of phosphorus in minerals, rocks, fertilizers, ferrous, and nonferrous metals. Since precipitation is performed in acid solution, the same procedure may be used when the separation of phosphorus from complex mixtures is to be performed prior to the analysis of the main cations.

Procedure

Dilute the nitric acid solution, containing boric acid and preferably not more than 50 mg of phosphorus to about 150 ml. Add about 10 ml of concentrated nitric acid and about 10 g of ammonium nitrate, heat to 50°C, add 20 ml of fresh molybdate solution for each estimated 10 mg of phosphorus present, stir with a mechanical or magnetic stirrer, let stand preferably for 12 h, filter through a fine porosity sintered glass filter or a fine porosity paper filter if the precipitate is to be redissolved. Wash with a cold 5% solution of ammonium nitrate, and twice with 5 ml of water containing about 1% of nitric acid. If the precipitate is not excessive and no interfering substances (Fe, Al, and Ti) are suspected, dry, and heat in oven for 30 min at 200°C, weigh (repeat twice) as $(NH_4)_3PO_4 \cdot 12MoO_3$.

Gravimetric factors:

$$P \quad = (NH_4)_3PO_4 \cdot 12MoO_3 \times 0.016507$$
$$P_2O_5 = (NH_4)_3PO_4 \cdot 12MoO_3 \times 0.037823$$

Usually it is better to dissolve the yellow precipitate in dilute ammonium hydroxide, acidify and reprecipitate before weighing, especially when the precipitation has been performed from solutions containing interfering and partially co-precipitating substances.

Ammonium molybdate reagent. Dissolve about 125 g of ammonium molybdate in about 500 ml of water, containing about 15 ml of ammonia and 50 g of ammonium nitrate. Pour this solution, stirring vigorously, into 1 l of a solution containing 400 ml of concentrated nitric acid. Wait if milky precipitate forms while stirring. Let stand overnight and filter. Keep in a polyethylene bottle for not over one month.

Precipitation of magnesium ammonium phosphate hexahydrate

The classical accurate method for gravimetric determination of phosphorus requires subsequent solution of the phosphomolybdate in diluted ammonia and reprecipitation as magnesium phosphate

Procedure

Dissolve the yellow precipitate in about 30 ml of a 5 N ammonium hydroxide solution containing 1 g of ammonium citrate. Wash the filter with hot water followed once by 5 N hydrochloric acid, and again by dilute ammonia. Add about 1 g of citric acid to the basic solution, boil and filter if milky. Acidify this solution with hydrochloric acid in the presence of methyl red as indicator, and add an excess of 10 ml of 5 N HCl, add about 10 ml of a 10% $MgCl_2 \cdot 6H_2O$ solution (100 g $MgCl_2 \cdot 6H_2O$ + 50 g NH_4Cl + 5 ml NH_4OH diluted to 1 l, filtered if necessary and just acidified with hydrochloric acid, and stored in a polyethylene bottle) and precipitate slowly by dilute ammonia drop by drop while stirring, and wait to let the precipitate settle between each addition of the base. When the color of methyl red changes, add 5 ml of concentrated ammonia to the solution, which should not exceed 200 ml in volume, and let stand for about 4 h, but better overnight. Filter and redissolve the precipitate caught on the filter, after decanting and washing three times with dilute ammonia, in about 30 ml of 5 N hydrochloric acid, collecting the solution into the same beaker in which the major part of the precipitate was decanted. Add 1 ml of magnesium chloride reagent to this solution which should not exceed 100 ml. Precipitate by dropwise addition of concentrated ammonia till first crystallization occurs. Switch to dilute ammonia immediately, and proceed the dropwise addition till the color change of the methyl red occurs, and no visible new precipitate forms. Then add an additional 5 ml of concentrated ammonia and let stand for 4 h or better overnight. Ignite the precipitate to pyrophosphate, $Mg_2P_2O_7$, with the same precautions as described under Magnesium.

Gravimetric factors:

$$P = Mg_2P_2O_7 \times 0.27831$$
$$P_2O_5 = Mg_2P_2O_7 \times 0.63776$$

Interference. The preceding procedure does eliminate the most serious interferences and co-precipitation, if performed correctly. The effect of fluorine on the solubility of the molybdophosphate is minimized by complexing it with boric acid and by expelling most of it first with nitric acid. The boric acid also protects the glassware from the attack of the hydrofluoric acid. The repeated evaporations with hydrochloric and other mineral acids remove and precipitate the silica, which would otherwise interfere.

The possible interference of vanadium, titanium, zirconium, iron, and aluminum, is kept to a minimum by the addition of citric acid to the solution from which magnesium ammonium phosphate is to be precipitated.

The following precautions are necessary. No sulfuric or hydrochloric acids should be added to the solutions from which phosphomolybdate is precipitated. Vanadium and arsenic contaminate the precipitate and interfere. Tungsten co-precipitates, but most of it is easily removed as yellow tungstic acid when hydrochloric and nitric acids are used to evaporate the dissolved sample to dryness (not in platinum crucible in this case), and if the acid-insoluble yellow residue is filtered.

Today, many laboratories prefer the colorimetric molybdivanadophosphoric complexing method.

CARBON

Carbon belongs to the most common trace elements in the igneous rocks, in which it is more abundant than phosphorus, manganese, sulfur or fluorine. In the cosmos, it belongs to the main elements. In carbonate rocks, such as limestone and dolomite, carbon is the main constituent. In siliceous rocks carbon occurs mainly in carbonate minerals but also structurally substituting as (CO_3)-groups in phosphates and silicate minerals. Native carbon is found in the form of graphite or diamond. Calcite, $CaCO_3$, is the most important among the mineral species containing carbon. Carbon is concentrated in the later stages of the crystallization of igneous rocks forming the volatile carbon dioxide, CO_2.

Carbon is the central building block of all organic matter and essential to all life. It can be said that a half million distinct organic compounds are presently known to man, whereas the total of inorganic derivatives of all the other elements is measured in only tens of thousands. The impact of the revolutionary development of the organic chemistry on our civilization is, therefore, enormous. Natural oil and coal, and organic materials derived from carbon, form the main basis for modern technology. The carbonate rocks, limestone, dolomite, and magnesite are the chief raw materials for cements, ceramics, refractories, and numerous other industrial products. Determination of carbon is routinely required in complete analysis of these materials.

Chemical properties. Carbon forms chains and rings in which the carbon atoms are linked together. The formation of chains of identical atoms is called *catenation*. In addition, carbon combines with either electropositive or electronegative elements just as easily. Both these properties cause the multitude of organic carbon compounds mentioned above. In this respect, carbon is unique among the elements.

Carbon is usually determined as carbon dioxide, CO_2, volumetrically or by absorption in soda-asbestos or soda-lime after volatilization with an acid and purification. In case graphite is present, as in some rocks, combustion in oxygen atmosphere or even direct oxidation by a sulfuric-chromic acid mixture may be employed, In steel, metals and alloys, a direct combustion to carbon dioxide in the presence of catalyzers such as copper oxide is generally used. In samples containing organic matter, direct combustion in controlled oxygen atmosphere, purification from other possible gases, and weighing as CO_2 is universally applied. Samples containing carbides, SiC, for example, are analyzed by combustion in the presence of oxidants such as Pb_3O_4.

The CO_2 gas precipitates white $CaCO_3$ or $BaCO_3$ from ammoniacal solutions of lime water, or calcium or barium chlorides. Observation of the effervescence caused by, or bubbles rising in a test tube upon addition of acids to sample, is the simplest qualitative test. The CO_2 gas is odorless and colorless. Even a slight effervescence of a rock or mineral powder with a drop of sulfuric or hydrochloric acid is a good indication of the possible presence of carbonates.

Simplified method

Applicable to rocks, minerals, ceramic materials, clays, raw materials for refractories, and cement.

Samples with less than 6% CO_2 present as carbonate, are best analyzed by the SHAPIRO and BRANNOCK (1962) method, developed at the United States Geological Survey. This method is a good example of simplification and speed without sacrifice in accuracy. It can be used in a series of specific cases where only the carbonate carbon is determined, and it has some advantages when compared to combustion, and chromic-sulfuric acid methods insofar as graphite carbon is unaffected and can be ignited and determined separately.

Procedure

Weigh 1, 0.5, or 0.25 g of sample powder, depending on whether it contains up to 1.5, 3, or 6% CO_2, respectively. Transfer the powder through a funnel into a glass tube with a sidearm (Fig.10), add 1 g $HgCl_2$ powder, add 2 ml of saturated $HgCl_2$ solution, add heavy paraffin oil, viscosity 335/350, to the mark, filling the side-

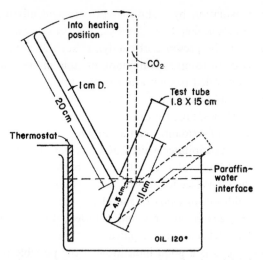

Fig.10. Tube for simplified CO_2 determination developed by SHAPIRO and BRANNOCK (1962) at the U.S. Geological Survey. After loading in upright position, the tube is tilted and lowered into a paraffin bath kept at a constant temperature. The gas-collecting arm is now in a vertical position in order to trap all gas bubbles, as indicated by the dotted image of the tube. (Courtesy of L. Shapiro and W. W. Brannock.)

arm, carefully removing all entrapped air bubbles. Place the side-arm in a vertical position with a clamp, and add 2 ml of 5 N HCl. Immerse the tube up to the oil–water interface in an oil bath heated at 120°C, and heat for 5 min. Remove and cool the tube and the sidearm with running tap water and measure the height of the gas column using a millimeter scale.

Repeat the procedure simultaneously with a known amount of a standard mixture containing preferably a similar amount of CO_2 and calculate:

$$\% \ CO_2 \text{ in sample} = \frac{\left(\begin{array}{c}\text{wt. of standard} \\ \text{in g}\end{array}\right) \times \left(\begin{array}{c}\% \ CO_2 \text{ in} \\ \text{standard}\end{array}\right) \times \left(\begin{array}{c}\text{mm length gas-} \\ \text{column of sample}\end{array}\right)}{\text{(sample wt. in g)} \times \left(\begin{array}{c}\text{mm length gas-} \\ \text{column of standard}\end{array}\right)} \qquad (6)$$

Remarks. Simultaneous treatment and cooling of sample and standard is important because of the volumetric nature of this method. $HgCl_2$ is added to prevent metallic iron from forming hydrogen. Identical amounts of sample powder of the same grain size are to be preferred for both the standard and the unknown because of the presence of air which can not be accounted for by this method. Pyrite is not affected by HCl, and pyrrhotite and chalcopyrite are attacked so slowly by the 5 *N* HCl that the 5-min heating period does not seem to be sufficient to cause significant errors by the formation of H_2S, but if considerable sulfides are present, the method fails. Up to 3 g of sample may be taken quite comfortably and the amount of acid slightly increased. Standards can be prepared by homogenizing a mixture of carbonate-free quartz with a known amount of dry, powdered, low alkali $CaCO_3$. After some experience the tube can be marked in percent if the same weight of sample is used.

Combination of classical acid decomposition and combustion methods

Acid attack. Applicable to samples with over 6% CO_2, such as carbonates, limestone, dolomite, magnesite, and their mixtures with silicates, also to samples where the simplified method fails because of interfering compounds. Large samples and samples of wide composition range can be used.

Combustion attack. Applicable to steel, graphite-bearing rocks and substances, and metallurgical products in general.

Procedure and equipment. In the following procedure carbon dioxide is freed from the sample either by combustion or by acid attack, purified, absorbed, and determined gravimetrically. The apparatus described permits the determination of carbon in a wide variety of substances and composition ranges. It can be suspended from a rod or kept by clamps in a desired positon.

The combustion train (Fig.11A, no.*1–9*)

    (*1*) Oxygen gas tank with double valve.

(2) Needle valve.

(3) Bubbler containing concentrated sulfuric acid to estimate the gas flow, and to absorb traces of SO_2, and water. This can be replaced simply by a rotameter gauge with a flow rate range of 30–300 ml/min if the oxygen gas is sufficiently dry and free of SO_2 and water.

(4) Tube packed with "anhydrone"-anhydrous magnesium perchlorate, $Mg(ClO_4)_2$, or calcium chloride, $CaCl_2$, to absorb moisture,

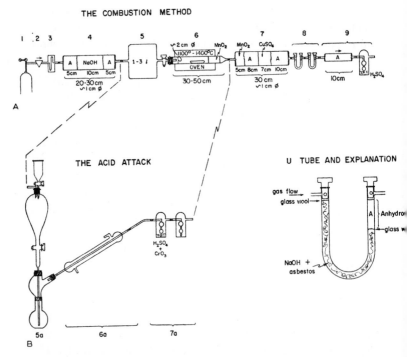

Fig.11. The combustion train for the determination of CO_2 with the necessary modifications for the acid attack method. Normally the combustion of sample is accomplished in the electric oven, procedure A; if the acid attack is preferred, parts 5 and 6 are substituted by parts 5a, 6a, and 7a. An expanded U-tube, loaded for absorbing the generated purified CO_2 gas, is shown in the lower right corner. The same combustion train with changes indicated on p.120 can also be used for accurate determination of total water. (For further description see text, pp.113–116.)

followed by ignited asbestos of 20–30-mesh size impregnated with sodium hydroxide, NaOH, to absorb carbon dioxide (commercially available CO_2 absorbant, preferably with a color indicator added so that the saturation of the absorbant may be noticed before any CO_2 is lost, may be used), followed again by "anhydrone" to absorb moisture released by the reaction:

$$CO_2 + 2NaOH \rightarrow Na_2CO_3 + H_2O$$

Each purifying reagent is separated by ignited glass wool or asbestos.

(5) Aspirator of 1–5 l capacity to equalize the gas flow and, if necessary, to supply quickly enough oxygen into the combustion chamber to prevent the formation of CO.

(6) Combustion tube of porcelain, mullite, or platinum, about 50 cm in length and an electric furnace that can be heated to at least 1,100°C, preferably to 1,400°C. The tube contains a combustion boat of porcelain, alundum, or zirconium oxide. Platinum can be used if metals and sulfides are absent, and if no CuO is mixed with the sample. At the end of the tube about 8 cm of manganese oxide, MnO_2, is loosely packed between glass wool plugs, and placed so that about 2–4 cm of it is outside the furnace.

MnO$_2$ reagent. Add 5 g $KMnO_4$ and 3 g $MnSO_4$ to 200 ml of a 1 N nitric acid solution. Mix vigorously for about 10 min and keep adding small amounts of $MnSO_4$ till the permanganate color nearly disappears. Wash the black MnO_2 precipitate well with water, compress and cut into small pieces about 1–5 mm in diameter, dry in a glass tube at 100°C while passing air or oxygen through, and finally heat for 30 min at about 300°C. Separate the fine powder using a sieve, and discard.

(7) Tube with anhydrous magnesium perchlorate for removal of moisture, and anhydrous copper sulfate on pumice in the center to react with H_2S and HCl. This reagent is prepared by soaking crushed ignited pumice in saturated $CuSO_4$ solution, draining and drying at 200°C, or making a paste of $CuSO_4$ and water, drying at 200°, breaking the cake, and retaining the coarse 10–20-mesh material.

If the sample contains *considerable sulfide*, another combustion tube is inserted between tubes 6 and 7 and heated at 400°C. This tube contains platinized silica gel separated by glass wool from an ironized asbestos layer, which is prepared by igniting asbestos impregnated by a saturated $Fe(NO_3)_3$ solution at 1,000°C.

(*8*) Two U-tubes, each containing 2/3 of its length of soda-asbestos to absorb CO_2, separated with glass wool from an anhydrone layer, second tube serving mainly as a guard. The size of the tubes may be varied according to the amount of CO_2 to be determined. These tubes are always kept tightly closed when not in use during testing or analysis.

(*9*) Bubbler with concentrated H_2SO_4 to protect from moisture and indicate the flow of gas. It is of advantage to insert as indicated another anhydrone tube before this bottle. This assures balanced atmospheric conditions in both U-tubes.

The acid-attack train (Fig.11B, no.*5a–7a*)

(*5a*) Decomposition flask, 100–250 ml with 50 ml tap funnel for addition of acid.

(*6a*) Condenser to prevent the hydrochloric acid from entering the sulfuric acid bubbler bottle. This may be omitted if concentrated phosphoric acid, H_3PO_4, is used for the attack.

(*7a*) $H_2SO_4 + CrO_3$ bubbler bottle and guard, to absorb SO_2 and moisture. Concentrated sulfuric acid is saturated with CrO_3. To absorb *chlorine*, one can use sulfuric acid diluted with an equal amount of water and saturated with silver sulfate, Ag_2SO_4.

Combustion procedure

Regulate the oxygen flow through the apparatus chain to about 50 ml per min or 2–4 bubbles/sec. Heat the combustion tube to 1,100°C, and after 10–15 min, weigh the first absorption tube (no. 8), and repeat till the weight of this tube remains constant.

Spread the sample over the alundum combustion boat (not platinum) carefully, and cover with ignited CuO, lead chromate, $PbCrO_4$, or another suitable flux. It is advisable to spread some CuO also under the sample. Place the boat into the hot combustion tube and connect immediately with the aspirator while oxygen is flowing at the same rate as during the calibration. Ignite for 10–15 min and permit the oxygen flow to continue for about 10 min afterward. Remove and clean the outside walls of the U-tubes, close them, permit to cool, open shortly to equalize pressure, close, and weigh. The second tube should show no or only slight weight increase. The weight difference, minus the blank which is always run with the same amounts of reagents and for approximately the same time, corresponds to the CO_2 in the sample, if present as carbonate. To convert to carbon, multiply by 0.27291.

Remarks. Platinum boats can be used without fluxes for rock and mineral samples containing no considerable sulfides and little graphite. In samples with graphite and carbonate, *carbonate carbon* is determined by acid treatment first. The *graphite* is separated and filtered off, after hydrofluoric acid decomposition and digestion of another sample in 5 N hydrochloric acid, washed, dried, weighed, mixed with CuO, and determined by combustion. In rock or mineral samples containing scapolites, cancrinite, or carbonate-apatite, total carbon can best be determined by the combustion method, mixing the sample with CuO or $PbCrO_4$.

Acid-attack procedure

Connect the decomposition flask *5a* with tube *4* and the sulfuric acid bubblers *7a* and tube *7*. Add to the flask 50 ml of acid, to be used for decomposition, and boil gently. Bubble oxygen through at a rate of 2–4 bubbles/sec for 10 min while boiling the acid gently, disconnect the U-tubes and weigh, repeat, checking all connections until constant weight is achieved. Remove the lower bottle, discard the acid, dry, and connect the bottle tightly to the decomposition apparatus.

Transfer 0.2–10 g of the sample powder, depending on the CO_2 content, into the dry decomposition bottle, moisten with 1–3 ml of boiled warm water added through the tap funnel, and first bubble oxygen through to displace the air. During this procedure, the bubbler bottles *7a* are kept disconnected from the condenser for about 1 min. With the whole apparatus chain tightly connected and oxygen bubbling through at a rate of 2–4 bubbles per sec, add 20–50 ml of 5 N hydrochloric acid through the top funnel drop by drop. When the effervescent reaction subsides, start heating slowly, and finally boil after about 10 min. During this procedure, observe carefully the rate of flow of the gases through the system, avoiding as far as possible a violent release of gas during the addition of acid. Remove both U-tubes, close them, wipe clean, cool, open to equalize pressure, close, and weigh. The difference of weight is CO_2, which is reported as such in carbonates.

Gravimetric factor:

$$C = CO_2 \times 0.27291$$

Remark. Instead of pressurized oxygen, one can also use air and suction at the other end of the train.

HYDROGEN AND WATER

Hydrogen is the most abundant element of the universe. In crustal rocks, hydrogen belongs to the minor constituents. About 71% of the earth's surface is under water, and about 10% of the earth's solid surface is covered with continental ice. The concentration of hydrogen in the form of its oxide into one layer, the hydrosphere, is characteristic for this element. With the exception of the inert gases, hydrogen combines with all elements. In addition, its affinity to carbon in organic compounds results in more hydrogen compounds than of any other element.

In rocks, minerals, and industrial raw materials derived from these, hydrogen is usually analyzed as water. Simplifying, water can be conveniently divided into the so-called *essential*, the *surface-adsorbed* or *hygroscopic*, and also the *zeolite* forms. The respective terms used in rock analysis are *combined or plus water*, H_2O+, for the essential, and *hygroscopic or minus water*, H_2O-, for the nonessential water. The scope of this book permits only treatment of these types of water from an analyst's standpoint. In order to interpret his results structurally, an analyst must, however, distinguish between the merely surface-adsorbed water, the occluded water, the

zeolite water which is fixed in position but unessential to the preser-
vation of the structure of the crystal, crystalline water present as
H_2O groups like in gypsum ($CaSO_4 \cdot 2H_2O$, for example), the hydroxide group (OH) important in clays, the "acid" hydrogen as
hydrogen ion (H^+), the hydride hydrogen of the metals, and finally
the adsorbed hydrogen on metal surfaces. In reporting results of
analyses the proper form of occurrence of hydrogen in the substance has to be considered.

Chemical properties. Hydrogen is structurally the simplest of all
atoms. Its nucleus, the proton, has a charge of $+1$. Only one extranuclear electron is present. The three hydrogen isotopes, hydrogen, ^1H, deuterium, ^2H, and tritium, ^3H, (H, D, T) are of im-
mense importance in the nuclear and thermonuclear sciences and
industry. The hydrogen molecule H_2 is the lightest gaseous molecule known. It is chemically not very reactive, but it is a strong
reducing agent.

The presence of water is rarely in doubt, and the problem is to
determine the mode or modes of occurrence of hydrogen in the substance to be analyzed. One can see from the preceeding introduction
that this often may become a major task when this element is fixed
differently in the same substance or mixture. In many instances
this task is simplified if one knows the minerals, but when clay
minerals and water of crystallization are present and in other special
cases, one has to use infrared absorption spectroscopy to identify
and estimate the approximate relative amounts of the molecular
water, the hydroxyl groups, and the extent of the hydrogen bonding.

The quantitative determination of hydrogen in solid samples is
generally performed by ignition and oxidation with oxygen and an
oxidizing flux. The resulting water is absorbed on a suitable medium
and weighed.

Total water

The *accurate combustion method* is applicable to a wide variety
of samples ranging from rocks to metals, and with a wide range of
hydrogen or water content.

The apparatus consists of a combustion train similar to that shown in the combustion method for carbon (Fig.11), with the following changes.

For *accurate* work, if small quantities of hydrogen are to be determined, add between gauge no.*3* and absorption tube no.*4*, and between the combustion tube no.*6* and U-tubes no.*8*, two additional 15 cm long platinum-asbestos filled combustion tubes in respective electric ovens to be heated at 400°–500°C. Remove tube no.*7* and fill both U-tubes, no.*8*, with only phosphorus pentoxide or anhydrous magnesium perchlorate mixed with some glass wool to prevent packing and to counteract deliquescence.

For *rocks, minerals, ceramics, and refractories and their raw materials*, replace the MnO_2 catalyzer in tube *6* with platinized asbestos, remove tube *7*, fill both U-tubes as above.

Procedure

Run tests with the same amount of lead chromate, lead oxide, or copper oxide fluxes as is to be introduced during the analysis. Heat at 900°C for rocks and steel. Some rock and mineral powders have to be fluxed at 1,100°–1,200°C. Check first for sublimation temperatures of the fluxes used and never heat above.

Introduce about 1 g of sample mixed and covered with the flux in the combustion boat into the hot furnace, connect the aspirator, open the U-tubes and bubble oxygen at a rate of 3–4 bubbles/sec through the system. Flux for about 20 min, disconnect the U-tubes, close them, cool, open to equalize pressure, and weigh as water.

Gravimetric factors:

$$H = H_2O \times 0.11190$$
$$H = O \times 0.12600$$

The Penfield method

Apparatus. Penfield tube, made from borosilicate tubing, about 30 cm long and with three bulbs, each 2–3 cm in diameter. One bulb situated at the end of the tube, and another two bulbs in

Fig.12. The Penfield tube.

mmediate succession starting at a distance of 10 cm therefrom (Fig.12).

Procedure

Mix 1 g of the fine sample with 2–3 grams of powdered lead dioxide, PbO_2, lead chromate, or anhydrous sodium tungstate. Use lead dioxide if minerals which are difficult to decompose, such as hornblende, are present, and a mixture of equal parts of lead dioxide and lead chromate if major carbonates are present. Use anhydrous sodium tungstate if sulfur is present. If fluorine is present as a major component, it is advisable to mix 3 g of lead dioxide with 1 g of ignited calcium oxide for a flux. These reagents are first fused separately in a platinum crucible at about 800°C, then crushed, and the fines sieved off. A complete blank run is necessary. Penfield tubes are dried at 110°C in an oven and kept in a desiccator.

Transfer the mixture with a special funnel with a long stem into the bottom part of the Penfield tube. Tap it and pass some of the plain reagent powder to "wash" the funnel and to cover the sample with a thin layer of the flux. Connect glass-to-glass the opening of the Penfield tube to a piece of glass tubing of the same diameter, with a capillary end. Use rubber tubing for this. Wrap the double bulbs and the upper 20 cm long end of the Penfield tube with wet cold cloth or wet paper. Hold the tube in a nearly horizontal position and heat the lower bulb, slowly raising the gas flame to the point when the bulb just starts contracting. Keep the flame at that level and flux for 10–30 min depending on the composition of the substance and the condition of the flux. Now raise the flame to full blast till the lower bulb contracts considerably, and use a torch flame and iron tongs to separate it by twisting from the upper part of the Penfield tube while holding the tube in a horizontal position. Cool and clean the tube walls and let stand for 30 min near the balance with the capillary still in position. Remove the capillary tube, once again wipe the Penfield tube with a dry cloth, especially where the rubber tubing was attached, (hold the tube with tongs) and weigh. The time interval between the detachment of the lower bulb and the weighing must be constant in all determinations. Note that no desiccator is used. The blank must be run with the same amounts of flux as used with the sample. After weighing, keep the tube at 110°C for 2 h, cool in the same place for 30 min and weigh again. The weight difference is total water from which the *hygroscopic moisture*, H_2O-, is subtracted, before reporting as H_2O+.

Remarks. This method has been considerably simplified by SHAPIRO and BRANNOCK (1962) in the United States Geological Survey laboratories by fusing in a narrow 15 cm long borosilicate test tube and placing a rolled pre-weighed cylinder of filter paper

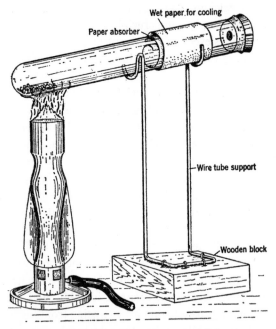

Fig.13. The modified Penfield method used at the U.S. Geological Survey. (Courtesy of L. Shapiro and W. W. Brannock.)

as absorber near the mouth of the tube (Fig.13). The upper part of the tube is placed in a perforated polyethylene beaker filled with crushed ice and some sodium chloride. Water condensed on the tube walls is carefully wiped off by the filter-paper cylinder which is weighed in a stoppered weighing bottle. When carefully timed and conducted, this method gives fully satisfactory results and takes only 10–20 min to perform.

Empirical correction factors may be applied to both methods if suitable standards are available, which unfortunately is seldom the case of total water. It is important to time all the individual steps, and to run exact blanks and checks.

Carbon dioxide interferes with both methods as does organic material. If these are present, one is advised to use the ignition method.

Procedure for nonessential water

Moisture adsorbed by the sample is determined by spreading 1 g of the sample powder in a platinum crucible and heating twice for 1 h at $105° \pm 5°C$ in an oven. The difference in weight is reported as minus water, H_2O-.

Remarks. It should not be assumed that the result represents only the hygroscopic water, because many compounds and minerals partly lose their water at a temperature below 100°C. A practical rule is to consider a constant loss of less than 0.5% at 105°C as due to moisture in a sample stored under normal room conditions. Inconsistent results with increasing losses or even gain after repeated timed oven treatment indicate the presence of substances such as clay minerals, zeolites, or water of crystallization that starts escaping below 105°C. In such cases, the total water figure should be used in theoretical calculations of the analysis. If the work is performed in relatively dry and stable atmosphere, as in some modern air-conditioned laboratories and in desert regions, and standardized grain size of the powders is used, one may use the empirical mean value of H_2O- in the normal type powders analyzed by the laboratory to estimate the hygroscopic water in samples which show irregular weight losses, assuming of course that hygroscopic or deliquescent substances are absent. In the determination of minus or plus water, meaningful and accurate results can only be obtained if the procedures are well standardized, including the grain size distribution of powders, the blank runs, and the careful timing of successive steps and weighings.

FLUORINE

Fluorine belongs to the most abundant trace elements in the lithosphere. In most siliceous rocks it is detected in amounts from 0.2 to 0.005% and in phosphate rocks in amounts from 0.1 to about 4%. In minerals the negative F^- ion has practically the same space requirements as oxygen, O^{2-}, or the hydroxyl group, OH^-. Thus is explained the occurrence of fluorine in most silicate minerals containing (OH)-groups. Especially micas and amphiboles contain fluorine. Fluorite, CaF_2, is the most important fluorine mineral in nature, and also in industry as a fluorine source and a fluxing agent in ceramic materials. Apatite, $Ca_5[(F, Cl, OH) (PO_4)_3]$, is the main carrier of fluorine in most rocks.

Today, fluorine is one of the most important elements in modern industry. The elemental fluorine is used as a missile propellant. Organic fluorine compounds are so numerous that their study represents a branch of organic chemistry. The inert nature and low toxicity of organic fluorine derivatives, such as the chlorofluoro-carbons, permits their use as refrigerants, and aerosol bomb propellants. Polymerized fluorinated derivatives of hydrocarbons are exceptionally resistant to acids, bases and high temperatures, as for example polytetrafluoroethylene (teflon). The use of teflon in crucibles, beakers, and all kinds of laboratory ware has considerably simplified many procedures and eliminated, in many cases, the problems of contamination and corrosion. In most complete rock analyses the determination of fluorine is required.

Fluorine can also have disastrous effects in construction by reducing the strength of some cements. On the other hand, its specific fluxing properties are sometimes desired in ceramics or cements. This makes fluorine analysis in cement and ceramic raw materials often essential.

Chemical properties. Fluorine, element atomic no.9, heads the seventh main group (VII) in the periodic system. Members of this group, fluorine, chlorine, bromine, and iodine, are called halogens. This collective name, as derived from the Greek, means "salt-builder". These elements are all highly reactive and therefore are not found in nature in the elemental state. They are only one electron short of the noble gas configuration, readily forming uni-negative ions or single covalent bonds. The non-metallic nature of the halogens characterizes their chemistry. The compounds formed between halogens and the light metals are typical "salts".

Fluorine is chemically the most reactive and the most electro-negative element. It stands between the nontransitional element oxygen and the noble gas neon. It forms ionic and covalent compounds. The ionic fluorides contain the F^- ion and resemble the oxides in their structures. Hydrogen fluoride, HF, is made by reacting sulfuric acid with fluorite. Its properties are very similar to those of water and it is an excellent solvent. In water solutions hydrofluoric acid is a relatively weak acid. Analytically, it is one of the most important acids, especially as a solvent and decomposition medium of siliceous materials. It forms a volatile silico-fluoride, SiF_4, which is used to expel silica in the classical gravimetric procedure. Calcium fluoride is poorly soluble in acids. The fluorine ion easily enters into different types of complexes, usually increasing the solubility of other ions. It interfers and partially inhibits the precipitation of aluminum hydroxide and ammonium phosphomolybdate. By forming an insoluble fluoride, the fluorine ion can also precipitate calcium earlier than desired during the classical analysis.

The precipitation of lead fluoro-chloride, PbFCl, and the precipitation of calcium fluoride, CaF_2, if large amounts of fluorine are present, have proven to be reliable *gravimetric methods*.

The detection of major fluorine is relatively simple when one uses the *etched glass* test:

The powdered sample is heated in a platinum crucible for 1 h with hot sulfuric acid under a polished wax-coated glass. Detection of corrosion or frosting effects over areas where the glass was not coated with wax reveals the presence of fluorine.

When small quantities of fluorine in the presence of silicates are to be detected, the so-called *hanging drop* test can be used: One mixes about 0.5 g of dry powdered sample with 0.1 g of powdered precipitated silica and a couple of milliliters of concentrated sulfuric acid in a test tube, and heats in boiling water for 30 min with loosely placed stopper into which a closed-end glass tube is inserted with a drop of water hanging from its tip. One fills the closed-end glass tube with some water for better cooling effect. A turbidity or a milky film on the drop of water indicates fluorine. This is caused by hydrated silica formed by hydrolysis of silicotetrafluoride: $2SiF_4 + 2H_2O \rightarrow SiO_2 + H_2SiF_6 + 2HF$.

Boron interferes with both tests, and some silicates are insufficiently attacked by sulfuric acid. In these cases, one fuses the powdered sample with 1/1 sodium–potassium carbonate mixture, extracts with water, filters, boils the solution for a short time with ammonium carbonate, filters, boils to 50 ml in a platinum dish, filters if necessary, adds calcium chloride solution while stirring, filters, and washes the precipitate with hot water. The calcium fluoride-carbonate precipitate can now be tested as above. Carbonate is first expelled by ignition if the "hanging drop" test is to be used because it may form a similar film.

Zirconium-alizarin test is suitable for trace amounts of fluorine, thus it is outside the scope of this text.

Decomposition and preparation of fluorine-bearing samples

Flux silicate samples with a tenfold amount of sodium-potassium carbonate (1/1), with an addition of about 1 g of silica powder in silica-poor or refractory samples. This constitutes the best attack on fluorine-bearing samples. Raise the temperature very slowly, watching that the mixture does not boil over. Do not fuse above dull red color of platinum crucible.

Carbonate rocks are first partially dissolved in dilute acetic acid without boiling, but stirring the mixture overnight magnetically in a covered beaker. The solution is then filtered and neutralized with sodium carbonate. The insoluble residue is fused in sodium–potassium carbonate mixture, and leached with water. Both neutral solutions are combined for precipitation of PbClF.

Fluorspar may be mixed with double its weight of quartz powder and also fused with the alkali carbonate mixture.

Phosphate samples are mixed with four times their weight of quartz powder and also fused with the alkali carbonate mixture. Leaching is always performed two or three times with as small quantities of water as possible, for example 20 ml + 10 ml + 10 ml. One collects the solution into a 100-ml volumetric flask.

In all these cases, contamination with silica and also phosphorus, if present, is unavoidable and for accurate results these have to be removed before proceeding with analysis for fluorine. If only silica is to be removed, add about 5 g of ammonium carbonate to 50 ml of the combined alkaline solution obtained through repeated leaching of the carbonate fusions with water, boil just a couple of minutes, permit to stand and cool for at least 2 h or overnight. Filter and wash the precipitate with cold 1% solution of ammonium carbonate, carefully neutralize with nitric followed by acetic acid drop by drop till color change of methyl red, and use the filtrate for the determination of fluorine by either of the methods described. In general, dissolution of samples in strong mineral acids is to be avoided and the decomposition by the carbonate fusion is to be preferred, if fluorine is to be determined.

If phosphorus is present in the sample in amounts over 0.5% it must be removed with zinc oxide before fluorine is precipitated. This procedure also removes silica, if still present, and should be used instead of the method described above if both silica and phosphorus are known to be present in the solution.

Acidify the cold combined leaching solution in a glass-covered Pyrex beaker by concentrated nitric acid drop by drop using methyl red as indicator. After the first color change, proceed acidifying by adding drop by drop of 5 N acetic acid, each time waiting till the effervescence caused by carbon dioxide subsides before proceeding. The pH of the solution should be 5.5, or as nearly as possible at the changing point of the indicator from red to yellow. Add about 1 g of solid zinc oxide powder, heat to incipient boiling, vigorously stirring, and cool at once in cold water for 30 min while stirring. Let the precipitate settle, decant the solution by suction through a fine porosity filter, rinse the precipitate three times and finally wash it on the filter a total of six to seven times with 2–3 ml of hot water. Try to keep the volume below 100 ml.

Procedure for gravimetric calcium fluoride (modified Berzelius method)

This method may be recommended if the sample contains over 1% fluorine or above 10 mg F in the solution. Add, to the solution from which silica and phosphorus have been removed, a few drops of 5 N nitric acid till just acid with methyl red. Add 5–10 ml of 10% calcium chloride solution and stir while adding drops of dilute ammonia till the color change of the indicator. Bring to incipient boiling while stirring, test whether enough CaCl reagent has been added, cool and filter through fine porosity paper. Wash seven to ten times with 5-ml portions of a cold solution containing 3% ammonium nitrate and 1% acetic acid, and finally four times with 5 ml of cold water. Dry and separate the precipitate from paper, burn the paper separately and ignite the precipitate in a platinum crucible at 600°–800°C twice for about 10 min, cool in a desiccator and weigh.

If co-precipitation of calcium carbonate is suspected or if sodium carbonate has been used as neutralizing and co-precipitating agent in addition to calcium chloride, treat the dried precipitate with 5 N acetic acid, evaporate to dryness, digest with 5 ml of water, filter and wash four to six times with 3 ml of hot water, dry, ignite separately from the paper, and weigh as CaF_2.

For check, convert the fluoride to sulfate by evaporating three times to fumes with a few drops of 5 N sulfuric acid. Cool, dilute each time with 2 ml of water, and finally fume with some added ammonium carbonate till the volatilization of the excess of the sulfuric acid. Ignite at 600°–800°C for 10 min twice, cool, and weigh as $CaSO_4$.

Gravimetric factors:

$$F = CaF_2 \times 0.48664$$
$$F = CaSO_4 \times 0.27911$$

Procedure for lead chlorofluoride method by Starck (modified)

This procedure may be recommended if fluorine is present as a minor constituent and the sample contains from 5–100 mg F. Using small volumes of liquid, even smaller amounts than 5 mg F have been satisfactorily handled by this author.

Add 1 ml of 5 N acetic acid to the nearly boiling neutral solution obtained after the zinc oxide or the ammonium carbonate treatments. Slowly add 50 ml of hot lead chloronitrate solution while stirring. Bring to boiling and immediately put the beaker aside and let cool overnight. Filter through a fine porosity sintered glass crucible and wash with cold saturated lead chlorofluoride solution. Use four 5-ml portions of this solution followed by four 5-ml portions of acetone. Dry at 150°–200°C, weigh as PbClF.

Gravimetric factors:

$$F = PbClF \times 0.07263$$
$$O = F \times 0.42107$$

Reagents. The lead chloronitrate solution is prepared by dissolving 10.5 g $PbCl_2$ and 13 g of $Pb(NO_3)_2$ by boiling in 1 l of water. Decant or filter before use if $PbCl_2$ crystals form.

The *wash solution* is prepared by precipitating PbClF as above from solution containing about 200 mg NaF with about 100 ml of the reagent described above. Wash well with cold water, stir, and boil the precipitate in 2 l of water, cool, and filter before use. Make sure some precipitate remains undissolved or crystallizes upon cooling.

Remark. Fluorine in amounts less than 2 mg can only be estimated by this method. Accurate procedure for trace amounts of fluorine is given in another volume. Fusing samples bigger than 2 g in an attempt to bring the total concentration of fluorine to a few milligrams, is not recommended because it usually results in too bulky residues and presents technical problems during the leaching and washing procedures.

SULFUR

In crustal rocks, sulfur stands sixth in abundance in the sequence Ti—H—P—Mn—F—S. This group represents the most abundant trace elements in the lithosphere. In meteorites, sulfur stands fifth among the most abundant elements, and it is estimated that in the bulk composition of the whole earth sulfur represents about 2.7 weight percent, which makes it the sixth most abundant element on this planet.

The main sulfur minerals are either sulfates or sulfides. Calcium sulfates gypsum, $CaSO_4 \cdot 2H_2O$, and anhydrite, $CaSO_4$, are widely distributed in sedimentary sulfate deposits. Sulfides pyrite, FeS_2, and pyrrhotite, FeS, and numerous other sulfide minerals are concentrated in so-called sulfide ores. Sulfur also occurs in native form in industrially important quantities. Sulfuric acid, elemental sulfur, sulfur dioxide, hydrogen sulfide, and their derivatives belong to the industrially most important compounds. Hydrogen sulfide, H_2S, and sulfur dioxide, SO_2, are gases which are widely distributed in nature.

In rocks, minerals, industrial inorganic raw materials, and their products, one usually analyzes for either sulfate or sulfide sulfur. Only these sulfur compounds will be discussed here, because otherwise it is impossible to condense the highly complex analytical chemistry of sulfur to fit the restricted nature of this book; consequently, analysis of sulfur in materials of organic origin such as oil, oil shale, lignite, and coal tar is not discussed here.

Chemical properties. Sulfur, element atomic no.16, belongs with oxygen, selenium, tellurium, and polonium to the main sixth (VI) group of nontransitional elements. It stands under oxygen and above selenium. In the third period it stands between phosphorus and chlorine. Elements of this group exhibit a pronounced non-metallic character. Most of their compounds are covalent. Sulfur has a tendency to catenation, representing one of the few elements resembling in this respect carbon.

Only the *sulfate ion* and the *sulfide sulfur* will be considered here. The chemist analyzing industrial raw materials is mostly concerned with these sulfur compounds, as is the geochemist, the mineralogist, and the geologist. In addition, the elemental sulfur, and sulfur dioxide and hydrogen sulfide, are of considerable interest to this group of scientists.

For the purpose of *gravimetric analysis* sulfur is usually transformed into the sulfate form and precipitated as the acid-insoluble barium sulfate. This universal procedure is known, however, to be unsatisfactory due to co-precipitation of most anions, occlusion of foreign matter, and partial decomposition of the precipitate into sulfuric acid and loss of such through volatilization during the ignition in the presence of absorbed hydrogen ions.

Strictly speaking, it is impossible to give a method generally applicable to most substances containing sulfur in the sulfate form. This is true to some degree in respect to all elements treated in this book. In no case is it more difficult, however, to suggest a satisfactory and reasonably simple gravimetric method applicable to a wide variety of materials, than in the case of sulfur. When determining small quantities of sulfur, one must bear in mind that the sulfate ion is one of the most common contaminants of all reagents used in a modern laboratory and that even the gas flame is a source of this element. Paradoxically, it appears to this author, that the sulfate ion is generally considered by beginners an "easy" one to determine gravimetrically.

The presence of sulfate is detected by the barium sulfate test and of sulfide by the typical odor of the evolving H_2S when sulfides are treated by acids. Hydrogen sulfide gas blackens moist lead acetate paper.

Gravimetric determination as barium sulfate

Samples containing gypsum or other acid-soluble sulfates may be dissolved in dilute hydrochloric acid. The insoluble residue will be mostly silica, but it may partly be barium sulfate in the sample, in which case it has to be treated separately by fusion with sodium carbonate.

Procedure after decomposition of sample by hydrochloric acid

Boil about 500 mg of fine-powdered sample for a few minutes in about 30 ml of 5 N HCl, dilute to about 150 ml, boil for another few minutes and filter hot through a medium porosity filter. Wash well with hot water till the volume is about 200 ml, heat to boiling, and add, while vigorously stirring, about 20 ml of a boiling hot 10% solution of $BaCl_2 \cdot 2H_2O$. Digest on water bath for 2 h, cool, and filter through the finest porosity paper. Wash with cold water, dry, separate the filter, ignite at 850°C separately, and weigh as $BaSO_4$.

Phosphates interfere if present. Carbonate rock powders with sulfides are first ignited for 10–15 min to oxidize sulfides.

Decomposition of sample. Rock and mineral samples containing calcium, strontium, barium and lead, some sulfates, and less than 1% sulfides, but no major chlorides nor phosphates, are best decomposed by fusion with a sixfold amount of dry, anhydrous, sulfate-free, sodium carbonate and some potassium nitrate as oxidizing medium. This procedure also removes iron, aluminum, magnesium, and calcium, which all may interfere later in the analysis if present.

Sodium carbonate flux. Put about 2 g of Na_2CO_3 into a 30-ml platinum crucible, add 1,000 mg of the sample, and about 300 mg of KNO_3, cover with 2 g of $NaCO_3$ and mix well, cover with about 1 g of $NaCO_3$ and ignite, slowly increasing the temperature, under platinum lid cover with the crucible inserted into an asbestos plate to avoid contamination with sulfur from the gas flame. When the effervescence of carbon dioxide subsides put the hot crucible into a vertical electric oven heated to 1,100°C, and keep there for 10–30 min, with the oven lid slightly displaced until the melt appears clear. Cool and transfer the clean crucible into a beaker, just cover with water, add 1 ml of ethanol, let the cake decompose, remove and wash the crucible and break up the lumps with glass rod. Decant and filter into a flask through fine porosity paper, and wash eight to ten times with a 1% Na_2CO_3 solution, stirring the residue up with the water jet each time. Reserve the residue for the determination of *barium*. Neutralize the filtrate through a short funnel with a bent tip to prevent spraying, adding concentrated hydrochloric acid drop by drop and using methyl red as indicator. Boil the just acid solution, adding diluted hydrochloric acid, if necessary, till the carbonates are expelled and the acidity is just at the changing point (red) of the indicator, then add 2 ml of 5 N HCl. If total sulfur is expected to be only a few tenths of one percent, the volume should be reduced to 50–100 ml. Some analysts suggest evaporating the acid solution to dryness with hydrochloric acid to destroy the nitrates and to filter the silica off before proceeding. This, however, may cause losses of sulfate by occlusion and absorption in gelatinous silica. The solution should now be 200 ml in volume.

Gravimetric procedure after fluxing with Na_2CO_3

Make another 200 ml of solution containing 0.5–1.0 g of $BaCl \cdot 2H_2O$ and 2 ml of 5 N HCl. Add some paper pulp, boil, and to this add, drop by drop from a tap funnel, the acidified filtrate of the carbonate fusion, while vigorously stirring with magnet. Add 2 ml of ethanol and digest for about 2 h on a water bath. Then cool, decant and filter, and refilter if necessary through the same fine porosity paper till filtrate is clear. Wash with cold water till free of chlorides, checking with silver nitrate. Dry, separate the precipitate from the paper preferably into another crucible, burn the paper separately, add the white ash to the main precipitate, and ignite with a partly open crucible at 850°C twice for 10 min. Depending on whether one wants to recalculate to sulfur, if mostly sulfides are present, or sulfate, the following factors are used:

$$S \quad = BaSO_4 \times 0.13737$$
$$SO_3 = BaSO_4 \times 0.34300$$

If silica is suspected, one can add to the dry, weighed, precipitate 1–2 drops of 5 N H_2SO_4 and about 2 ml of HF, evaporate to dryness and ignite twice at 850°C as before. This, of course, will transform barium, if occluded as chloride, to sulfate, causing positive errors.

Often in the actual work the combined errors cancel each other. An analyst must especially in this case mobilize all his knowledge, and skills, and take great care to avoid reporting contaminated precipitates of unknown constitution as $BaSO_4$. Often, fusing this precipitate with only about 0.5–1.0 g of $NaCO_3$, leaching as shown above, and reprecipitating in the presence of less sodium salts, may be required. In the case of minor sulfur, running a complete blank with the same quantities of reagents is an absolute requirement.

Decomposition of sulfide samples by aqua regia and bromine

Pyrite and other sulfide and sulfide ore samples should not be ground finer than 80–100 mesh because the sulfide sulfur is easily oxidized and volatilized as SO_2. This makes the decomposition of such samples a relatively lengthy procedure, and it is advisable to use only 200–300 mg samples, which should be preferably freshly ground.

Place 200–300 mg of the dry sample into a flask or a glass beaker, add 10–15 ml of carbon tetrachloride saturated with bromine through a funnel with a bent tip. Add 5 ml bromine, and 10 ml of concentrated nitric acid, cover with a glass and let stand overnight. Carefully add 10 ml of concentrated hydrochloric acid and let stand for one hour. Warm up slowly under glass till bromine color and nitric acid fumes are expelled.

If any black sulfide particles remain, cool, and add 5 ml of bromine-saturated carbon tetrachloride, 2–3 ml of bromine, 15 ml of HNO_3, and 5 ml of HCl. Let stand covered for one hour, then heat on the water bath till the reaction subsides.

Gravimetric procedure after the decomposition of sulfides

Add twice 5–10 ml of HCl and evaporate each time to dryness, moisten with 2 ml of concentrated hydrochloric acid, turn the flask till all of the crust is wetted by the acid, add 50–70 ml of hot water, bring to boiling, let cool, and while still warm but not hot, add 100–200 mg of mossy or fine granular zinc (which is less messy to handle than powdered aluminum) and shake till the color due to the Fe^{3+} ion disappears, cool, filter, and wash ten to fifteen times with cold water till the volume is near 400 ml, add 4 ml of 5 N hydrochloric acid, mix gently, and add 15 ml of 10% barium chloride solution drop by drop without stirring, then stir for a few minutes, and let the precipitate settle for 2–4 h. Filter as above by decanting through a fine porosity filter, wash with small quantities of water, dry, burn the paper carefully and ignite at 850°C. To speed the procedure, it is advantageous to use a sintered silica crucible and to apply suction. The precipitate can then be ignited directly in the crucible.

Remark. In the presence of major phosphates the methods described here do not give a phosphate-free barium sulfate precipitate. It is therefore important to *test* the *ignited precipitate* after the ignition *for phosphate.*

Fuse with 0.5–1 g of sodium carbonate, leach, acidify the filtrates with nitric acid, evaporate to dryness, moisten with nitric acid, add water, and filter off the silica if present, and precipitate as described under phosphorus. If the barium sulfate has *not* been moistened by sulfuric and evaporated with hydrofluoric acid to sulfur trioxide fumes, as advised above in the case of contamination with silica, all phosphorus originally present in the ignited precipitate mainly as barium pyrophosphate, $Ba_2P_2O_7$, is retained, and can be recalculated as such and subtracted from the total weight of the barium precipitate.

This procedure takes less time than the separation of phosphorus before the barium sulfate precipitation as magnesium ammonium phosphate, which also introduces considerable magnesium and ammonium salts which interfere in the precipitation of barium sulfate.

The zinc oxide–sodium carbonate fusion

The combined acid bromine attack decomposes satisfactorily most sulfide ore mixtures. When lead, phosphorus, arsenic, barium, strontium, and much calcium are present in sulfide ores, one fuses about 0.5 g of the powder with eight times its weight of zinc oxide–sodium carbonate mixture, which is prepared by mixing one part of zinc oxide with two parts of sodium carbonate. The packing and fusion are performed as described under sodium carbonate fusion. The leaching and washing are done with hot water, and the residue is discarded after boiling twice with about 20 ml of water. The combined filtrates are acidified and boiled free of carbon dioxide, using methyl red as indicator, and an excess of 4 ml of 5 N hydrochloric acid, then diluted to about 400 ml, boiled, and $BaSO_4$ precipitated by adding dropwise 20–30 ml of a hot 10% barium chloride solution. The precipitate is treated as described above.

BARIUM

Barium, belongs to the most abundant trace elements of the earth's crust. In many rocks, and especially ores, also in industrial raw materials, carbonate rocks and minerals, ceramics and cements, it is present in amounts that can be handled *gravimetrically* with an advantage in accuracy, simplicity, and reliability.

The main aspects of the chemistry of this element are described under Calcium. This short section is placed after sulfur because it is usually more convenient to determine barium and sulfur after the sodium carbonate–potassium nitrate fusion and the leaching process which separates these elements.

Calcium, barium, strontium, and magnesium, and zinc if not removed by hydrogen sulfide, are present in the combined filtrates and washings after the removal of silica and the double or triple precipitation of the R_2O_3 group.

Barium is best determined from these filtrates if no sulfates are present or have been added through the hydrogen sulfide separation of zinc or other metals, in which case most barium is already co-precipitated with the R_2O_3 group.

Procedure

Transfer the water-leached *residue* of the sodium carbonate fusion from an opened filter paper by means of a hot water jet into a beaker, and wash the paper with hot water. Bring the solution up to 75 ml and add, while vigorously stirring, 4 ml of 1/1 sulfuric acid and 0.1 g of sodium sulfite, Na_2SO_3. Stir magnetically on water bath for 1 h, let cool, and filter through a fine porosity paper, wash three times with 5–10 ml of cold water, dry in a platinum crucible, and burn the filter paper off carefully till the precipitate is white, add 1 g of sodium carbonate, slowly increase the temperature and fuse finally for 10 min at full blast. Cool and decompose the melt in 25 ml of water. Filter through a fine porosity paper, and wash the fine precipitate with a 1% sodium carbonate solution. Dissolve the precipitate from the filter paper into a beaker with a 0.1 N hydrochloric acid solution and wash the paper ten times with water. Discard the undissolved part of the precipitate which is mostly silica. Dilute the solution to 100 ml and add 3 ml of 1/1 sulfuric acid, stir, and digest overnight. Filter through a fine porosity paper, and wash five times with 5 ml of cold water. Dry, and burn off the paper carefully in a platinum crucible as described above, and weigh after ignition at 850°C. Moisten with a drop of 5 N sulfuric acid, evaporate, ignite again, cool, and weigh as $BaSO_4$. Strontium follows barium in this procedure. Test also for lead and calcium.

Gravimetric factors:

$$Ba = BaSO_4 \times 0.58845$$
$$BaO = BaSO_4 \times 0.65700$$

Remarks. This procedure has been used successfully by PECK (1964) of the United States Geological Survey. It is designed for small quantities of barium, but by increasing the volumes and the amount of sodium carbonate, up to 1% BaO can be determined. This covers well the usual composition range in rocks, ceramics, and cement raw materials.

Substances rich in barium are decomposed also by the carbonate fusion and the precipitation performed from large volumes with dilute sulfuric acid; the sulfate precipitate is better re-fused at least one time with sodium carbonate.

CHLORINE

Chlorine belongs to the most abundant trace elements in crustal rocks. It is abundant in all natural waters, and the sea water contains nearly 2% chlorine. The most important chlorine product is hydrochloric acid, which together with sulfuric acid form the basis of modern chemical industry. Because of this, just as in the case of the sulfuric acid, most chemical reagents are contaminated to some extent with chlorine.

Chlorine is determined routinely in water, brines, sea water, clays, soils, biotite, hornblende, bricks, glasses, bone ashes, in sulfate, carbonate and phosphate rocks, some ores and rocks, cinders, bleaching powders, and it is a constituent of many minerals.

In some cements, calcium chloride is added as activator and sodium chloride as accelerator. When the granulation of concrete is done with sea water the final product will contain chlorine. Due to chlorine's corrosive effect on steel, analyses of chlorine in some concrete may be required.

The main chlorine minerals are halite, $NaCl$, sylvite, KCl, carnallite, $MgCl_2 \cdot KCl \cdot 6H_2O$, kainite, $KMg(Cl/SO_4) \cdot 3H_2O$, and bischofite, $MgCl_2 \cdot 6H_2O$, all common in salt deposits.

Chemical properties. General chemical properties of halogens were discussed in the section on fluorine. Chlorine is less reactive than fluorine. Chlorides of alkaline earth metals are water-soluble, whereas silver chloride is insoluble in dilute acids. In this, chlorine differs from fluorine, and in general there is more coherence between Cl, Br, and I, than between these elements and F.

The insolubility of silver chloride is widely used in *gravimetric determination* of chlorine and silver. The same property is used to detect chlorine. Silver chloride is slowly reduced by light with separation of dark metallic silver. The solutions from which silver chloride has been precipitated show a typical opalescence. Both these properties make the detection of minute quantities of chlorine relatively simple. Only silver bromides and iodides show similar characteristics. All silver halogenides are easily soluble in ammonia.

For analytical purposes, three types of chlorine occurring essentially as chloride are distinguished: the water-soluble, the acid-soluble, and the total chlorine in solids by carbonate fusion. The sample preparation is performed accordingly.

Sample decomposition

Water-soluble chlorides. Comprised mostly of alkali and alkaline earth chlorides, and also some rocks which may be contaminated by adsorbed sodium chloride.

Stir 500–1,000 mg of the sample powder for a few minutes in 100 ml of water, let settle, if solution is not clear, add 0.5 g of sodium nitrate, mix, let settle, filter, cool, wash the precipitate with water containing sodium nitrate, acidify with five drops of nitric acid, and use the filtrate for chlorine determination.

Acid-soluble chlorides. Boil 500 mg of the pulverized sample for 2–3 min in 100 ml of nitric acid solution containing 20 ml of 5 N HNO_3. If carbonates are present, make sure that the solution remains acid. Keep the beaker covered with a watch glass during the procedure. Filter and wash the insoluble residue with hot water containing a few drops of nitric acid. Use the filtrate for chlorine determination.

Carbonate fusion. Applicable in general to samples of complex composition and where interferences in the determination of chlorine by precipitation as AgCl are expected.

Mix the sample powder with three to five times its weight of 1/1 sodium-potassium carbonate, reserving some carbonate to cover the mixed powder before fuming. Fuse as described under Silicon. Leach and wash the residue well with 1% sodium carbonate solution, boil finally with sodium carbonate and filter. Acidify the filtrate under watch glass with 5 N nitric acid drop by drop using methyl orange as indicator, cover and let stand for 8–10 h. If flocculent silica does not precipitate, use the filtrate for chlorine determination after a careful expulsion of carbonates under a watch glass on water bath.

Caution. Ammonium hydroxide precipitation treatment of each of these three solutions is necessary if much iron, or other heavy metals which may reduce silver, are present, or if silica has precipitated after the carbonate fusion. In such cases, add about 3 ml of concentrated nitric acid, stir the cold solution, and carefully neutralize by ammonia, boil, filter, and wash the hydroxide precipitate with warm 1% ammonium nitrate solution made just ammoniacal, using methyl red as indicator. Acidify the filtrates with a few drops of nitric acid. The volumes should not exceed 200 ml

Gravimetric method for chlorine as silver chloride, AgCl

Procedure

(To be performed in a brown glass flask and in subdued light.) Add to the clear cold nitric acid solution of the sample, which should not exceed 200 ml, 5 ml of silver nitrate solution, while stirring. The silver nitrate solution is prepared by dissolving 85 g of $AgNO_3$ in 1 l of water containing 1 ml of concentrated nitric acid. Permit the precipitate to settle, add a few drops of the reagent, and observe whether more precipitate forms. If this is so, add a few milliliters of the reagent and repeat the check until no new precipitate can be observed. Heat the solution to incipient boiling, while vigorously stirring with a magnetic stirrer. Disconnect the heating elements as soon as boiling starts and let cool on the plate while continuing the stirring. Let the precipitate coagulate, preferably in a dark cabinet or box. When the solution is cold, discontinue stirring, and test with a few drops of $AgNO_3$ solution to check whether all chlorine is precipitated. If this is so, filter through a fine porosity sintered glass crucible by decanting, and wash with cold water which contains 2–3 drops of concentrated nitric acid per 100 ml. Test the filtrate with dilute hydrochloric acid until no silver can be detected. Dry the precipitate first at 110°C till it appears dry, increasing the oven temperature to 150°C.

Note. Before being used for filtration, the sintered glass or a Gooch crucible has to be treated in succession with ammonia, water, nitric acid and water, and kept at 150°C before weighing to standardize its weight.

If the precipitate looks dark, moisten with one drop of the wash solution followed by one drop of 5 N hydrochloric acid and dry before weighing as AgCl. If, during the analysis procedure, contamination was indicated or is suspected due to the composition of the sample, the silver chloride may be redissolved in dilute ammonia and reprecipitated under better controlled acidity conditions.

This method is not applicable to trace amounts of chlorine, frequent in geological samples.

Gravimetric factors:

$$Cl = AgCl \times 0.24737$$
$$Ag = AgCl \times 0.75263$$

ZIRCONIUM

Zirconium, element atomic no.40, belongs to the major trace elements of the crustal rocks and is widely distributed, especially among the siliceous differentiation products of magmatic crystallization. The chief zirconium minerals are zircon, $Zr(SiO_4)$, and baddeleyite, ZrO_2. Hafnium partially replaces zirconium in its minerals, and, due to similar chemical properties, it always coprecipitates with zirconium when classical methods of separation are used. Zirconium is included in this book, due to its increasing use in refractories, and as cladding and armor in the cores of reactors. It is also used in metallurgical products and enamels.

Chemical properties. Zirconium belongs to the fourth group of the periodic system and its chemistry resembles that of titanium, hafnium, and thorium. It heads the elements of the second transition series, standing between yttrium and niobium. The similarity of its atomic size with hafnium is due to the so-called "lanthanide contraction". With this is meant the abnormal diminution of the atomic and ionic sizes with the increase in atomic numbers of elements from $Z = 57$ to $Z = 71$. This results from the filling by electrons of the 4f or other inner orbitals instead of the 6s orbitals.

Positive detection of zirconium is difficult by classical methods. If considerable zirconium is present, a white precipitate forms upon addition of a concentrated ammonium or sodium orthophosphate solution to a solution of the sample containing about 10 vol. % of concentrated sulfuric acid and hydrogen peroxide.

The insolubility of zirconium phosphate in strongly acid solution is also used in *gravimetric determination* of this element.

Decomposition of samples

Zirconium minerals and zirconia refractories are generally difficult to decompose completely and therefore the analyst will often have to try different attacks depending on the sample. Many will require prolonged fusion or two or three successive fusions with different reagents.

One of the most effective ways to fuse zirconium-bearing materials is the mixed *carbonate-boric acid fusion*, which may be recommended despite the inconvenience caused by the necessity of removing the boric acid with methanol before proceeding with the analysis.

Fluxing procedure

Mix two parts of anhydrous sodium carbonate and one part of fused and ground boric acid, B_2O_3. Mix about 8 g of this mixture with 0.5 g of impalpable sample powder in a platinum crucible. Fuse, raising the temperature slowly. When strong reaction ceases, proceed at full blast of a Meker burner or in a muffle furnace at 1,200°C for 20–30 min. Cool, decompose first in 100 ml of water, then add 50 ml of 5 N HCl, evaporate to hydrochloric acid fumes and add 30 ml of methanol. Mix well using a glass rod for stirring, and evaporate to sugary consistency till HCl fumes are barely detectable. Cool, add 20 ml of methanol, mix, evaporate, add again 20 ml methanol and 10 ml of concentrated hydrochloric acid to the cold residue, mix, and evaporate to complete dryness on a water bath.

Silica determination in zirconium-bearing materials

Procedure

Moisten the salts with a few milliliters of concentrated hydrochloric acid, add 50 ml of water, and permit the residue to disintegrate. Filter through a medium porosity paper and wash the residue with hot dilute hydrochloric acid solution ($\sim 1\%$ HCl). Save the filtrate and ignite the residue again, and, after weighing, volatilize silica with hydrofluoric acid, ignite the residue, and weigh. Fuse the residue with potassium bisulfate, dissolve, and add to the main filtrate.

Evaporate the combined filtrates in the same container as before to about 20 ml, add 10 ml of concentrated hydrochloric acid and 20 ml of methanol, mix well, and evaporate to a syrupy consistency on water bath. Dilute with water and transfer into a beaker, watching that the volume does not exceed 300 ml.

Precipitate the *ammonia group* as advised under aluminum, dissolve in dilute hydrochloric acid, and remembering to oxidize with bromine water, reprecipitate once or twice, depending on the amount of calcium and magnesium present. Ignite the combined oxides at 1,100°–1,200°C, and weigh if the total is desired. Fuse for 30 min with 3–5 g of potassium bisulfate, cool, add a few drops of concentrated sulfuric acid, and fuse again at 900°C, cool, and decompose in 60 ml of $5N$ sulfuric acid and 50 ml of water by boiling the solution for 5–10 min to decompose pyrosulfate. Evaporate to sulfuric acid fumes. Cool, pour this solution into 100 ml of cold water, rinse the beaker, digest for a few minutes on water bath, filter and wash the "residual silica" precipitate with hot water. Use the *combined filtrate* to determine zirconium as stated below.

Ignite the *residual silica* at 1,100°–1,200°C, weigh and volatilize with hydrofluoric acid, adding a few drops of 5 N sulfuric acid. Ignite the eventually remaining insoluble residue and weigh. The weight difference is the *total residual silica*. Add this to the main silica.

Fuse the residue with a few grains of potassium bisulfate, dissolve in 10 ml of 5 N H_2SO_4, and add the dissolved fusion product to the *combined filtrate* containing the R_2O_3 group. Transfer the solution into a 250-ml flask. Add 15 ml of concentrated H_2SO_4, cool to room temperature, and dilute to the mark. From this solution iron, aluminum, titanium, and zirconium are determined.

Zirconium as phosphate

Procedure

Take an aliquot of the solution of the R_2O_3 group, depending on the amount of zirconium present. If less than 1% of zirconium is present in the sample and only 0.5 g of sample was used, use the whole solution without addition of more sulfuric acid. Add 2 ml of 30% hydrogen peroxide solution, mix, and let the titanium peroxide color develop on water bath for 10 min. Remove from water bath and add, depending on the amount of zirconium, 5–20 ml of a warm, just prepared 10% ammonium phosphate, $(NH_4)_2HPO_4$, solution containing 10% of conc. sulfuric acid. The above solution is added drop by drop while stirring and the precipitate digested, placing the beaker on a dish filled with sand that is standing on a water bath. Let stand for 2–4 h, or overnight if precipitate is very small.

Filter through a medium porosity filter and wash ten times with about 20 ml of a cold neutral solution containing 5% ammonium nitrate, checking for the presence of the sulfate ion and stirring the gelatinous precipitate each time. As soon as only traces of sulfate can be detected, discontinue the washing. Dry the precipitate. Ignite the paper with great care and finally when it is completely decomposed, ignite for one hour at 1,200°C, and weigh as ZrP_2O_7.

Gravimetric factor:

$$ZrO_2 = ZrP_2O_7 \times 0.46468$$

In *accurate work* fuse this precipitate with sodium carbonate, leach with water as advised under Phosphorus, wash a few times with 1% sodium carbonate solution, dry, ignite, and flux with sodium carbonate the second time, leach, wash the insoluble residue with 1% sodium carbonate solution, ignite the residue, fuse it with potassium bisulfate, dissolve, precipitate with ammonia as advised under Aluminum, wash a few times with 1% ammonium nitrate solution made barely ammoniacal, redissolve and reprecipitate again, filter and wash eight to ten times with ammoniacal 1% ammonium nitrate solution, and ignite for 1 h at 1,200°C as ZrO_2.

Note. An acid insoluble residue may indicate zirconium phosphate.

Zirconium as mandelate

The method of Kumins is an example of recent important advances in gravimetry. Enough experience seems to have been accumulated by now to be able to recommend this method as a highly selective and relatively simple procedure, remarkably free from interference.

Procedure

Use an aliquot of the combined filtrates of the R_2O_3-group to precipitate the hydroxide with ammonia. Wash in the usual manner and dissolve in about 35 ml of hot 5 N hydrochloric acid, dilute to about 50 ml, and heat over hot but not boiling water bath. Add, while stirring, about 40 ml of a 20% para-bromomandelic acid, digest at the same temperature for about 20 min and filter through a medium porosity paper, wash a few times with a hot solution containing 1 N hydrochloric acid and bromomandelic acid, then two to three times with distilled water. Dry, and ignite at 1,000°C in a platinum crucible after burning the paper off carefully, weigh as ZrO_2.

Gravimetric factor:

$$Zr = ZrO_2 \times 0.74030$$

CHROMIUM

Chromium belongs to the major trace elements in the lithosphere. In igneous rocks, it is concentrated in the early differentiates, which are also enriched in iron and magnesium. In sedimentary rocks, it follows ferric iron and aluminum because of the similarity of the chemistry of these three elements. In analysis of siliceous materials chromium is a rare constituent and its determination is rarely called for but it is included here due to its abundance in some high refractory materials, especially magnesite refractories. Also some aluminum-rich clays, bauxite, shales, and schists, and different types of iron ores may be relatively rich in chromium. The main chromium ore mineral is chromium spinel or chromite, $FeCr_2O_4$, and its isomorphic mixtures.

Chemical properties. Chromium, element atomic No.24, is a typical transitional element exhibiting variable oxidation states. Its compounds show typical colorations. Chromium (III) is chemically similar to ferric iron and aluminum. Chromium (III) is the most stable and analytically important oxidation state of the element. The green color is typical to the Cr^{3+} ion. In hydroxide precipitates, it is often obscured by the reddish-brown color of the Fe^{3+} ion. Chromium (III) also forms numerous complexes.

The aqueous chemistry of the yellow and orange colored chromium (VI) compounds, especially the chromate, CrO_4^{2-}, and the dichromate, $Cr_2O_7^{2-}$, ions is analytically of great importance. The insolubility of lead, barium, and silver chromates, and the reduction of the chromate by ferrous iron are used, for example, in the determination of these elements. Chromic oxide, CrO_3, is used as a strong oxidizing agent.

Chromium, when present as chromate, may be precipitated as barium chromate, $BaCrO_4$, or reduced by a known amount of ferrous ammonium sulfate, the excess of which is then titrated back by permanganate.

The coloration of chromium compounds is a good indication, and the insolubility of the Pb, Ba, and Ag chromates is used for detection. When the ammonium hydroxide precipitate is boiled with sodium peroxide the filtrate is colored yellow in the presence of chromium, which is oxidized to form the chromate ion.

Decomposition of samples

Samples containing any of the *spinel type* chromium and iron minerals are generally difficult to grind and to decompose. Grinding to a fine impalpable powder in hardened-steel grinders, prolonged fluxing, and occasional refluxing of undecomposed particles, are essential procedures in this case.

To dissolve *metallic chromium* or chromium alloys, one should use dilute hydrochloric or dilute sulfuric acids and avoid the use of the nitric acid or aqua regia, which may prolong or inhibit the complete solution of the metal. To *separate* chromium from iron, the acid solution may be neutralized with ammonium carbonate, boiled with a few milliliters of hydrogen peroxide and filtered, washed with 1% $(NH_4)_2CO_3$ solution and the procedure repeated if a large iron hydroxide precipitate is present. Chromium ends up as chromate in the filtrate.

Siliceous samples containing less than 2% Cr_2O_3 can usually be decomposed and chromium oxidizied to Cr^{6+} satisfactorily by a sodium carbonate-potassium nitrate fusion. This flux may be tried also with samples containing considerably more chromium, and will often prove sufficient for complete decomposition, but not necessarily for complete oxidation of chromium to the hexavalent state.

Carbonate–nitrate fusion

Note. Platinum crucible is strongly attacked if the packing procedure is not followed.

Mix anhydrous sodium carbonate with one tenth of its weight of potassium nitrate, KNO_3. Mix about 1 g of finely-ground dry sample powder with a sixfold amount of the flux. Cover the bottom of the platinum crucible with about 1 g of dry pure sodium carbonate powder, add the mixture, and cover with 1–2 g of pure sodium carbonate powder. Fuse, slowly raising the temperature till the flux appears clear, cool the crucible for about 3 min, and quench placing into a 250 ml beaker with just enough water to cover the crucible. Leach till the crucible can be removed, washed and cleaned. Add 2–3 ml of ethanol to reduce the manganese, and decompose the lumps with a glass rod. Filter through a coarse porosity filter paper, wash seven to ten times with 1% solution of Na_2CO_3, boil with 5 ml of a 30% solution of H_2O_2 for 20 min, and reserve the filtrate for the determination of chromium.

This procedure separates chromium from iron, considerably simplifying the following determination of chromium. If considerable chromium and iron are present, and there is reason to suspect that not all of the chromium has been oxidized to chromate, transfer the iron hydroxide precipitate into a 250-ml beaker, add 100 ml water, 1 g sodium carbonate, and 5 ml of 30% hydrogen peroxide solution, boil for 10–20 min till hydrogen peroxide is expelled, filter, and wash with 1% sodium carbonate solution. Combine this filtrate with the first filtrate and reserve for the determination of chromium as chromate.

Sodium peroxide fusion for chromite ores

Fuse 0.5 g of a fine-ground, dry sample with 5 g of sodium peroxide in an iron or nickel crucible. Before fusion, mix the sample in the crucible with the yellow peroxide and cover with about 1 g of pure peroxide. Let the cake decompose in cold water, add 1 g of sodium peroxide and boil for 20–30 min to destroy the peroxide. The volume should be about 200 ml. Add about 10 g of ammonium carbonate, mix, boil, and filter through a coarse porosity filter. Wash the precipitate with hot 1% sodium carbonate solution. Reserve the filtrate for the determination of chromium as chromate.

Remarks. Keep in mind that the same oxidizing agents reduce chromates in acid solutions. The procedures given above have to be repeated two or three times if much chromium and iron are present. It is important to test for chromium before discarding the iron hydroxide precipitate. This is done by boiling with hydrogen or sodium peroxide and checking for the yellow color of the filtrate. For this purpose, the precipitate may first be dissolved in dilute nitric acid, neutralized with ammonium carbonate, and boiled with the peroxide. If yellow color appears, the filtrate must be added to the combined filtrates after boiling off the peroxide.

If high accuracy is not required, one may separate chromium from the remaining aluminum by ammonium hydroxide precipitation of the latter, as described under Aluminum, followed by reduction of the chromium in the filtrate and precipitation as hydroxide with a slight excess of ammonia, and an ignition to Cr_2O_3. This procedure is seldom used, however, because of the tendency of Cr_2O_3 to oxidize further during the ignition, giving in general higher ($\sim +1\%$) results than the theoretical. The gravimetric method consisting of precipitation as barium chromate is also now avoided after thermogravimetric studies have shown that $BaCrO_4$ is only stable below 60°C. Besides, the frequent presence of the sulfate ion often makes this method unapplicable. The most frequently used method involves the reduction of the chromate by an excess of ferrous ammonium sulfate, followed by back titration with permanganate or chromate.

Titrimetric procedure for the determination of chromium

Acidify the peroxide-free filtrates, which should be about 200 ml in volume, with 5 N sulfuric acid and add about 20 ml of the acid in excess. The chromate color changes from yellow to reddish-brown near the neutralization point, so that addition of indicators, which may reduce the chromate or the permanganate, can thus be avoided. Add 2–4 ml of concentrated phosphoric acid and stir. Add 20 mg of manganese sulfate, $MnSO_4 \cdot H_2O$, and a weighed excess of granular ferrous ammonium sulfate hexahydrate, $Fe(NH_4)_2(SO_4)_2 \cdot 6H_2O$ (14.24% Fe), to the cool solution, dissolve by stirring, and titrate immediately back with a 0.1 N solution of potassium permanganate, waiting for a few minutes to assure that the endpoint has been achieved.

If the peroxide has not been entirely removed from the combined filtrates and partial reduction of the chromate in acid solution is suspected, add 10–20 ml of a 1% solution of silver nitrate and 10–20 ml of a fresh 10% solution of ammonium persulfate to the acidified solution. Boil for 15 min, cool, and proceed as above. For exact work, blanks are run with the same amount of the ferrous salt, and particularly watching the endpoint which is somewhat difficult to see. In general, the green color of the Cr^{3+} ion changes into just perceptible violet with slight darkening at the end point upon the drop by drop addition of the permanganate.

The diphenyl carbazide method is suitable for trace chromium, and thus is outside the scope of this book.

BORON

Boron is a typical trace element. In crustal rocks, the average amount of boron is only a few p.p.m. This is in contrast to the other trace and minor elements treated in this book, which elements occur in crustal rocks in amounts varying from thousands to tens of p.p.m., $\sim 1,000$ to ~ 10 p.p.m. Boron is included here because it is frequently present in significant amounts in minerals, rocks, glasses, ores, fluxes, fertilizers, soils, plant ashes, waters, alloys, reactor materials, and numerous industrial products. Geochemically, boron is a lithophile element with a strongly pronounced tendency to combine with oxygen. During the process of magmatic crystallization it becomes enriched in the final or pegmatitic stages and also in volcanic emanations, which is due to its small ionic size and the volatility of its compounds. In siliceous igneous rocks, tourmaline is the main boron mineral, containing about 10% boric oxide, B_2O_3. In mafic rocks, axinite with about 6% B_2O_3 is the most common boron mineral. In evaporate sediments, large concentrations of boron are found in the form of such ore minerals as colemanite, $Ca_2B_6O_{11} \cdot 5H_2O$, borax, $Na_2B_4O_7 \cdot 10H_2O$, kernite, $Na_2B_4O_7 \cdot 4H_2O$, and ulexite, $NaCaB_5O_9 \cdot 8H_2O$.

Chemical properties. Boron, element atomic no.5, belongs to the main third group (III), and stands in the second period between beryllium and carbon in the periodic system of elements.

It is chemically different from its congeners aluminum, gallium, indium, thallium, and also from scandium, yttrium, and lanthanum, which are all metallic in their properties. While also being trivalent, it displays a strong tendency to form covalent bonds, and its compounds resemble those of silicon more than those of its congener aluminum. It is also a nonmetal in the chemical sense, forming a weak acid, $B(OH)_3$. Boron does not display amphoteric properties as aluminum does. Its chemical similarity to silicon is further demonstrated by its halides which hydrolyze easily. Borates resemble silicates in their structures and properties by being difficult to crystallize, and they also form glasses easily. The hydrides of boron are flammable gases which hydrolyze on contact with water.

Boron displays the greatest affinity to combine with fluorine and oxygen. Hydrolysis of boron compounds usually produces boric acid, $B(OH)_3$, which is monobasic and polymerizes in concentrated solution. Due to polymerization, the anionic structures of borates are cyclic or linear, which means that borate formulas as such give little stoichiometric information without steric interpretation. The complexity of the chemistry of boron greatly resembles that of silicon.

The increase of the strength of the boric acid when complexed by the addition of such polyhydroxy compounds as mannitol or glycerol is used in the *titrimetric determination* of boric acid in its solutions. Titrimetric determination is performed by first neutralizing the strong mineral acids till the color change of methyl orange or another suitable indicator. When mannitol is added, the "released" excess acidity of the mannito-boric acid can be titrated with standard alkali solution.

The mannitol or mannite formula is:

$$\begin{array}{ccccccc} H & H & H & OH & OH & H \\ | & | & | & | & | & | \\ HO\text{-}C\text{-}C&\text{---}C&\text{----}C&\text{---}C&\text{---}C\text{-}OH, & \text{abbreviated } [=C(OH)\cdot C(OH)=]. \\ | & | & | & | & | & | \\ H & OH & OH & H & H & H \end{array}$$

The reaction can be represented then by equation:

$$[=C(OH)\cdot C(OH)=] + B(OH)_3 + [=C(OH)\cdot C(OH)=] \rightarrow$$
$$[\equiv(CO)_2 \vdots\vdots B \vdots\vdots (OC)_2 \equiv] H + 3H_2O$$

An old proven and relatively simple *gravimetric method*, developed by Gooch, consists of distillation of boron as volatile methyl ester and its fixation by a suspension of calcium oxide and weighing as calcium metaborate after ignition. The precision and speed of the titrimetric procedures have made, however, the gravimetric procedure somewhat outdated. Naturally there are numerous other useful methods, which can not be mentioned here due to the restricted scope of this book.

Qualitative detection of boron is performed by heating the fine sample powder with sulfuric acid and methanol in a narrow test tube or a porcelain dish. When the alcohol fumes escaping the tube or the alcohol in the dish are ignited, a green flame indicates boron. A *bead test* in a loop of a platinum wire with a mixture of calcium fluoride and potassium sulfate with the sample powder, and ignited in the colorless part of the gas flame, also produces green flame, but in the presence of copper and barium this test is unreliable.

Decomposition of the sample

Most borates and borate-bearing carbonates are best dissolved in boiling dilute hydrochloric acid using a *reflux condenser.* For boron-bearing alloys, add some hydrogen peroxide. Flux acid-insoluble silicates and glasses in a platinum crucible with five to eight times of the sample weight of sodium carbonate, and leach out the resulting sodium borates under cover by hot water, filter, wash, trying to keep the volume small, neutralize with hydrochloric acid, add a few drops in excess using methyl red as indicator, and boil for 10–15 min to expel carbon dioxide, using a reflux condenser. If much iron, aluminum, and silica are present, heat the leached residue on water bath in a beaker with sufficient 5 N hydrochloric acid under cover till carbonates are destroyed. Cool and filter immediately, wash and add fine sodium carbonate powder just to neutralize the acid, filter and wash the iron, aluminum, and silica precipitate with warm water. Combine the filtrate with the main leachings, neutralize with 5 N hydrochloric acid using methyl red, and expel CO_2 using reflux condenser as above. In general, the volatility of boron compounds must be considered during all manipulations, and the use of borosilicate glassware should be avoided in very accurate work. Boron-free reagents must be assured.

Titrimetric procedure

Neutralize about 200 ml of the slightly acid solution, obtained as above or by distillation with methanol, using methyl red, with a 0.1–0.5 N standard alkali solution till color changes to yellow. Add

5–15 g of solid neutral mannitol, depending on the amount of boron present, stir and add 1 ml of 1% phenolphthalein in a 50% ethanol solution. Titrate with the standard alkali solution till the appearance of permanent pink color. It is advisable to run a test with a known amount of boric acid to determine the eventual losses or imperfections in procedure. 1 ml of 0.1 N NaOH = 0.003481 g B_2O_3.

Reagent. 0.1 N NaOH is best prepared as described on p. 179.

Gravimetric method

The apparatus consists of a stoppered flask connected through a reflux condenser to a tilted tube-shaped distillation vessel with a stoppered funnel at the other end. The distillation vessel is immersed in a paraffin bath heated at about 130°C (Fig.14).

Transfer 50–100 ml of the neutralized solution of the sample, prepared as described above, and brought up to a known volume but containing not more than 100 mg B_2O_3, through the funnel into the distillation vessel. Add a few drops of acetic acid. Close

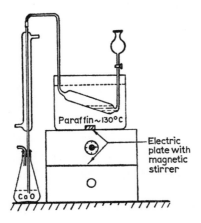

Fig.14. Distillation vessel for the gravimetric determination of boron.

the stopcock, and boil in a paraffin bath at 130°C. Distil over into the flask which contains a known amount of dry, freshly ignited CaO, preferably about 1 g. When all liquid is distilled over, cool the vessel, add 10–20 ml methanol and distil again to dryness, repeating the procedure three times. Add a few milliliters of acetic acid to the dry residue and again distil three times with about 100 ml of methanol. Disconnect the flask which contains the distillates and shake well, using a rubber stopper. Let stand for a few hours, transfer the emulsion, and evaporate in a large platinum dish on water bath, not permitting the alcohol to boil. Heat to destroy acetates and to make the residue more amenable to transfer into a crucible to be ignited. The material that remains on the walls of the flask and on the dish is dissolved in a few drops of 5 N nitric acid, transferred into a separate platinum or porcelain crucible and also ignited. The ignition of both crucibles is performed over a flame with caution, not above a dull red color of platinum, repeating till a constant weight is achieved. The total addition of weight to the known amount of CaO corresponds to the B_2O_3 in the sample aliquot used. A test should also be performed with a known amount of borate added. Phosphate may interfere.

Gravimetric factor:

$$B = B_2O_3 \times 0.31057$$

LEAD

Lead belongs to the more abundant trace elements of the earth's crust, being present in amounts of about 15 p.p.m. in crustal rocks. Lead is included here because it is one of the major constituents of sulfide ores, together with copper and zinc. Its affinity to sulfur is thus demonstrated. The main lead minerals are galena, PbS; cerussite, $PbCO_3$; pyromorphite, $Pb_5[Cl/(PO_4)_3]$; and anglesite, $Pb(SO_4)$. Numerous complex sulfides of lead, belonging to the so-called sulfosalts, are common in sulfide ores, but generally occur in smaller quantities. Lead is one of the most important heavy metals and raw products in industry.

Chemical properties. Lead, atomic number 82, is the most metallic of the group-IV elements. The electronegative character of these elements and the tendency to catenation decrease in the direction of higher atomic number: $C \to Si \to Ge \to Sn \to Pb$.

Lead is only weakly amphoteric, forming plumbates. The oxidation state of II is more dominant in lead compounds than the oxidation state of IV, which dominates the chemistry of lead's congeners.

The *separation* of lead as lead sulfate, $PbSO_4$, and subsequent gravimetric determination by weighing as such, may be regarded as the most important and well proven analytical method of almost universal nature.

Qualitative detection of major lead in its minerals or ores is performed by the same lead sulfate method. It is easily applicable if relatively large quantities of the metal are present. Potassium chromate precipitates yellow lead chromate, $PbCrO_4$, in weak acids, and hydrochloric acid in cold concentrated solutions precipitates white shiny needles of lead chloride, $PbCl_2$. Black lead sulfide, PbS, is precipitated in neutral or faintly acid solutions. Other sulfide group elements interfere.

Decomposition of samples

About 500–1,000 mg samples are preferably used. Most ores, alloys, carbonates, phosphates, oxides, and even lead-bearing sulfides, are best decomposed by heating the fine powders to boiling with 5 N or 1/1 nitric acid, and if decomposition appears incomplete, adding some hydrochloric acid or hydrogen peroxide to dissolve lead peroxide. In any case, after the removal of silica, the nitric acid treatment is followed by evaporation twice with sulfuric acid to white sulfuric oxide fumes.

Barium- and calcium-bearing samples and insoluble silicates and slags, or at least their insoluble residues obtained after nitric acid treatment, have to be fused with sodium carbonate, leached with water, and the insoluble residue treated with HNO_3. Proceed as under silica, filtering the silica off, and precipitating lead as sulfide from neutralized solution, which leaves barium and calcium in solution. Silica may also be removed by HF + H_2SO_4 acid treatment. Since the sulfuric acid treatment also renders a part of the possibly present antimony and tin salts insoluble, if these are present, a sodium carbonate fusion is also advisable.

Procedure, gravimetric major lead as PbSO₄

To the impure, fumed lead sulfate precipitate in concentrated sulfuric acid, add carefully, after cooling, about ten times its volume of cold water, digest for 10 min, filter and wash with a 1% H_2SO_4 solution saturated with lead sulfate, followed twice by ethyl alcohol to remove the acid. Dissolve the lead sulfate by vigorously shaking in 60–100 ml of a hot saturated ammonium acetate solution to which about 3 ml of concentrated acetic acid has been added per 100 ml. This is done by using 10–20 ml portions of this solution consecutively in an extraction-like procedure until no lead can be detected in the last portion. Evaporate to dryness and digest with a 10% per volume H_2SO_4 solution. Let stand, filter through sintered glass crucible of medium porosity, wash as above under suction, dry at 110°C, and ignite at 600°C.

Gravimetric factors:

$$Pb = PbSO_4 \times 0.68324$$
$$PbO = PbSO_4 \times 0.73600$$

Interferences with this method are many, but if the dissolution of sample is performed as described above and followed by acetate extraction, barium and silicon do not interfere. Ca, W, Nb, Ta, Bi, and Ag may still interfere and, if present in considerable amounts in the sample, must be considered. Precipitation of lead sulfide from a tartaric acid solution will separate lead from Ba, Ca, Nb, Ta, Bi, and Ag. Calcium is separated from lead sulfate precipitate by dissolving the lead salt in NaOH and filtering. Calcium sulfate is insoluble in this base. Tungsten may be dissolved by treating the precipitate with ammonia and ammonium tartrate mixture.

Procedure, gravimetric major lead as PbCrO₄

This method is used for more accurate lead determination follow-ing the acetate extraction described above. Make sure that the acetate solution is acid by adding 3–5 ml of concentrated acetic acid. Boil, and add 5–10 ml of 5% potassium dichromate, $K_2Cr_2O_7$, solution, while stirring until the heavy yellow precipitate changes to orange-red. Digest on water bath for 10–20 min, filter, wash with hot water, and wash twice with ethanol. Dry at 110°C to constant weight.

Gravimetric factors:

$$Pb = PbCrO_4 \times 0.64108$$
$$PbO = PbCrO_4 \times 0.69061$$

Interferences are Ba, Bi, Sb, and Ag, but barium will generally be removed by the acetate extraction. Bismuth can be kept in solution by adding 1–2 g of solid citric acid to the solution after the precipi-tation of lead chromate and stirring vigorously before filtering, but this may dissolve some lead chromate and should only be done if considerable bismuth is known to be present, and not at all if the precipitate is small.

There are numerous other good gravimetric procedures for lead which can not be mentioned here.

Titrimetric method

In routine work the *titrimetric procedure* is most frequently used. It is based on the reaction:

$$2PbCrO_4 + 6KI + 16HCl \rightarrow$$
$$2CrCl_3 + 2PbCl_2 + 6KCl + 8H_2O + 3I_2$$

from which we can see that Pb = 3I.

Procedure

Precipitate lead chromate as above and dissolve on the filter in 50–100 ml of a cold saturated sodium chloride–hydrochloric acid solution followed by cold water till the chromate color entirely disappears from the filter. Add about 1 g of potassium iodide, stir gently, and titrate with standard thiosulfate till the color of the liberated iodine fades, then add starch and proceed titrating carefully till the blue color of the starch changes to a greenish color.

It is of advantage to save a small portion of the chromate solution, and after rapid titration to the greenish color of starch, to add this solution to the main part and to proceed carefully to the exact endpoint. This is particularly advisable if much lead is present. In general, strong iodine solutions are apt to lose some iodine through evaporation and should be diluted before the addition of iodide. Repeat the whole procedure using a known amount of metallic lead dissolved in boiling concentrated sulfuric acid.

Reagent. Acid-saturated sodium chloride solution is prepared by adding 500 ml of 5 N HCl to 1 l of saturated sodium chloride solution, and filtering if necessary.

Titration of the lead chromate by ferrous ammonium sulfate

Procedure

Dissolve the lead chromate precipitate in 50–100 ml of the acid–saturated sodium chloride solution, which is prepared as stated above. Bring the volume, if necessary, up to about 100 ml and add 20 ml of a concentrated 1/1 sulfuric-phosphoric acid mixture. Add a few drops of diphenylamine (which is prepared by dissolving 1 g of the indicator in 100 ml of concentrated sulfuric acid), then add a known amount of the ferrous ammonium sulfate hexahydrate salt in solution in excess till the color of the solution changes to green, and titrate back with a 0.1 N potassium dichromate solution till blue color appears.

Before subtracting the excess volume corresponding to the titrated excess of the ferrous salt solution, remember to convert the back titrated volume on the basis of the relative normalities of both solutions by multiplying the milliliters by the ratio of normalities of both solutions. Calculate the amount of lead present on the basis of the spent ferrous salt solution:

1 ml of 0.1 N $Fe(NH_4)_2(SO_4)_2 \cdot 6H_2O$ = 0.006906 g Pb.

Interferences. In both titration methods, barium, if present in the sample in large amounts, is the most common interference. It may cause the retention of considerable lead when the sulfate precipitate is treated. When this is the case, the residue of the acid-acetate treatment, described under the lead sulfate method is evaporated to dryness with 5–15 ml of concentrated HCl. The residue formed is treated again with the acid acetate mixture by boiling and filtering. This filtrate should contain most of the lead retained during the initial treatment. It is added to the main acid acetate extraction solution.

NICKEL

Nickel is one of the more abundant trace elements in the crustal rocks, comprising about 80 p.p.m. thereof. Belonging to the iron family, it is concentrated in iron- and magnesium-rich rocks. The earth's core is supposed to consist of iron-nickel. The metal phase of meteorites is exceptionally rich in nickel, and a strong affinity of nickel to sulfur in terrestrial rocks has been demonstrated. For "sulfide" nickel ores, the combination Ni–As–Sb is typical. Cobalt usually follows nickel in its minerals but is more lithophile. However, both of these elements tend to occur in silicates, nickel replacing notably magnesium, in magnesium and ferromagnesian minerals. Weathering of nickel-bearing rocks often leaves it concentrated in hydrolyzate sediments forming hydrosilicates. Commercially, garnierite, $(NiMg)_6[(OH)_6/Si_4O_{11}]\cdot H_2O$, is of primary importance due to massive deposits in areas of weathered ultrabasic and serpentine rocks. Industrially, nickel is one of the most important metals.

Chemical properties. In the ordinary chemistry of nickel, element atomic no.28, one needs to consider only the bivalent oxidation state. Belonging to the first transition series and the eighth subgroup (VIII) of elements in the periodic system, it has a typical metallic character and displays a strong affinity to complex formation. Its chemistry is very similar to that of iron, but it displays no amphoteric tendency. Nickel metal is passivated like iron by strong nitric acid and does not dissolve in this reagent.

Quantitative determination of nickel is mainly based on its precipitation by *diacetyldioxime* (dimethylglyoxime) or electrolytic deposition as metal.

The scarlet crystalline precipitate obtained by addition of diacetyldioxime alcohol solution to the ammoniacal sample solution, containing tartaric acid to complex iron, is a good selective and sensitive *indication* for nickel.

Decomposition of samples

Strongly oxidizing reagents are used for decomposition of sulfides and arsenides of nickel, in most cases an acid treatment is, however, sufficient to dissolve all nickel compounds present. After the hydrochloric or nitric acid treatment, the silica is removed by dehydration and the solution used for the precipitation of nickel diacetyldioxime.

In *difficult cases* try to decompose the sample by mixing it with a three-fold amount of potassium chlorate and carefully adding 30–50 ml of concentrated nitric acid under cover. Use perchlorate hood! Place the covered dish on water bath and mix occasionally with a glass rod until the decomposition appears complete. Evaporate to dryness, add 10–20 ml of concentrated HCl, and dehydrate silica as above. Moisten with HCl, wash the sides of the dish with about 20 ml of hot water, boil up, filter, and wash the silica precipitate with hot water to which a few drops of HCl have been added.

If all these methods fail to decompose the sample and to render all of the nickel soluble, fuse with an eightfold amount of potassium bisulfate, extract the melt and filter off the silica. Sodium or potassium carbonate with about 200–500 mg of potassium nitrate may also be used for fusion.

Dissolve *metallic nickel* and its alloys in *dilute nitric acid* or even aqua regia, followed by a few milliliters of sulfuric acid, after which evaporate to fumes, and filter off insolubles.

Procedure, diacetyldioxime (dimethylglyoxime) method

Sample should preferably contain not more than 50 mg of nickel. Boil the hydrochloric acid solution of the sample with a few milliliters of concentrated nitric acid to oxidize the ferrous iron. Bring the volume to about 200 ml, add about 3 g of tartaric acid and 1 g of ammonium chloride and dissolve the salts by stirring. Carefully add concentrated ammonia till weak ammonia smell persists and follow by adding 3–5 ml of ammonia in excess, and put on water bath. If cobalt is present, add about 10 ml of 30% hydrogen peroxide. After 20 min, when the solution is warm, add about 40 ml of a 1% ethanolic diacetyldioxime solution and stir. Cool, filter through a medium porosity filter or sintered glass filter and wash with neutralized 1–3% ammonium nitrate solution. If the precipitate is slimy, dissolve it in hot 5 *N* nitric acid and reprecipitate without the addition of peroxide. Dissolve the precipitate trough the filter with hot 5 *N* nitric acid, preferably alternating the acid with hot water several times. See that the total volume is approximately 250 ml, make just ammoniacal and add 5 ml of concentrated ammonia in excess. Add no peroxide or tartaric acid. Put on water bath and proceed as above. Now use not more than 25 ml of the alcoholic 1% diacetyldioxime reagent, trying to adjust the amount of the reagent to about 5 ml per 10 mg of nickel in the sample. Filter warm and wash the precipitate with a little hot water six to eight times on a sintered glass crucible. Dry at 110°C for 2 h and weigh.

Gravimetric factors:

$$Ni = Ni[(CH_3)_2(CNO)_2H]_2 \times 0.20319$$
$$NiO = Ni[(CH_3)_2(CNO)_2H]_2 \times 0.25857$$

Interferences. Ferrous iron forms a similar compound but the nitric acid treatment oxidizes it to the three-valent state. The interference of Fe^{3+}, Al, Cr^{3+}, Mn, Ti, Zr, Pb, and Sn, is overcome by the addition of tartaric acid. Cobalt and copper in large amounts interfere. Copper is best removed as sulfide from an acid solution. Double precipitation of the nickel complex also minimizes the interference of copper considerably. The interference of major cobalt is best eliminated by treating the solution with potassium cyanide and hydrogen peroxide and an addition of formaldehyde before the precipitation. However, the addition of a few milliliters of hydrogen peroxide to the ammoniacal solution as described above will oxidize cobalt sufficiently in most cases so that the double precipitate will be pure enough.

COPPER

Copper is one of the more abundant trace elements in crustal rocks, averaging about 45 p.p.m. It is a typically chalcophile element, being concentrated in the sulfide phases during crystallization. It is also abundant in the metal phase of meteorites. Sulfides: chalcopyrite, $CuFeS_2$; bornite, Cu_5FeS_4; chalcocite, Cu_2S; and covellite, CuS; antimony- and arsenic-bearing sulfides: tetrahedrite, Cu_3SbS_{3-4}; enargite, Cu_3AsS_4—the so-called sulfosalts, and also some silicates like chrysocolla, $CuSiO_3 \cdot nH_2O$, carbonates like malachite, $Cu_2[(OH)_2/CO_3]$, or azurite, $Cu_3[OH/CO_3]_2$, and sulfates like chalcanthite, $Cu(SO_4) \cdot 5H_2O$, for example, are the most common copper minerals. Copper oxide, cuprite, Cu_2O, and native copper are also relatively common. In most sulfide ores, the determination of copper is required. Copper is usually the main constituent sought in the economic sense. Due to its high conductivity to electricity, resistance to oxidation and alloying properties, it is considered to be, after iron, the most important industrial metal. Copper is included here for these reasons.

Chemical properties. Copper, element atomic no.29, has one s-electron outside the completed d-shell, but otherwise it has little in common with alkalies. It displays three oxidation states, I, II, III. The cuprous (I) and the cupric (II) compounds are analytically of importance. Copper (I) compounds are unstable. Copper, due to its typically metallic character, and the tendency to form complex ionic solutions, is classed as a transition element, together with its congeners in the first sub-group – silver and gold. Chemically, copper differs considerably from these elements.

Applying classical methods, copper is precipitated from acid solution (which is preferably 3–4 N H$_2$SO$_4$), together with the so-called hydrogen sulfide group (Pb, As, Sb, Sn, Ag, Cd, Hg, Bi, and Mo). This separates copper from iron, aluminum, and zinc. Further treatment of the precipitate is necessary to separate copper. Copper can also be separated by precipitating it on such metals as aluminum, zinc, cadmium, or iron, but the simplest and most accurate method is to *electrolyze* and weigh as copper. This method is especially recommended because it introduces no contamination in the dissolved sample.

The appearance of blue color when a copper-bearing solution is made ammoniacal is the best *indication*. The blue color is due to complexes of the type $[Cu(NH_3)(H_2O)_5]^{2+}...(Cu(NH_3)_4(H_2O)_2)^{2+}$. This reaction is, however, not very sensitive and nickel gives a similar color. When one uses "amine" complexes instead, such as the ethylenediamine for example, the coloration is much more intense. Many organic reagents have been developed to detect small quantities of copper, such as the ammonium α-nitroso phenylhydroxylamine (cupferron), for example. These have also been used successfully for *quantitative* determination of this element. However, the precipitates are non-stoichiometric, slimy, and therefore, often difficult to wash, and the pH ranges required are narrow, with too many interferences to cope with to recommend them for a beginner.

Decomposition of copper-bearing ores or copper metal or alloys is mostly effected by nitric and sulfuric acids, aqua regia, or if necessary, a potassium bisulfate fusion. Hydrobromic acid is effective on antimony-, arsenic-, and tin-bearing alloys; hydrofluoric acid on refractory materials and silicates. Preliminary sodium hydroxide attack followed by nitric acid is recommended for aluminum alloys.

Electro-gravimetric procedure

Decompose a 1-g sample powder in a 1/1 mixture of 5 N sulfuric and nitric acids. Evaporate to fumes and evaporate further to syrupy consistency until salts start crystallizing on the beaker bottom, filter the silica off after diluting with water, and wash with dilute sulfuric acid (if silica precipitate is colored, fuse with potassium bisulfate to liberate the remaining copper). Add 25 ml of 5 N H_2SO_4 and 25 ml of 5 N HNO_3, dilute to about 200 ml, boil, and let cool. Electrolyze for 4 h in a covered beaker at 4.0 A, stirring the solution continuously. If there is time, electrolysis should be performed without stirring at 1.0 A for about twenty hours. See that the beaker sides and the cover are washed clean into the beaker a few hours before the completion. Weigh the platinum cathode (electrode) after washing with water and alcohol, and drying.

Note. Test the electrolyzed solution, adding first citric acid and then ammonia in excess and a 0.1% solution of diethyldithiocarbamate. If brown color develops, indicating incomplete deposition, the remaining copper must be removed by extraction with carbon tetrachloride before discarding the solution or proceeding with further determinations.

Silver precipitates with copper and if present must be determined separately. It is advantageous to separate silver as chloride before the electrolysis. Major tin interferes by precipitating as metastannic acid when the acid solution is boiled to remove nitrous oxide.

Gravimetric factors:

$$CuO = Cu \times 1.2518$$
$$Cu_2O = Cu \times 1.1259$$

ZINC

Zinc belongs to the more common trace elements. Its average concentration in the lithosphere is in the order of 60–70 p.p.m. It mostly occurs as the sulfide sphalerite, ZnS, but the carbonate smithsonite, $ZnCO_3$, and the silicate willemite, $Zn_2(SiO_4)$, are also common zinc minerals. Zinc and cadmium are closely associated geochemically, being concentrated mostly in the late hydrothermal sulfide phase of crystallization. Zinc is of major importance as an alloy metal, especially in brass. Laboratory glassware sometimes contains this metal. Zinc is included in this part of the book be cause of its frequent presence in sulfide ores as a major constituent.

Chemical properties. Zinc, element atomic no.30, cadmium, and mercury, belong to the second sub-group (II) of the periodic system. They do not display multiple valence and are regarded as non-transitional elements, but standing between the groups Cu–Ag–Au and Ga–In–Tl, they still display a tendency to complex formation. Zinc hydroxide, $Zn(OH)_2$, shows amphoteric characteristics dissolving in alkalies to give zincates.

Analytically, zinc is separated after the removal of the hydrogen sulfide group which consists of Cu, Pb, As, Sb, Bi, and Cd. This is done by precipitating with hydrogen sulfide from a hot 3 N sulfuric acid solution containing only the sulfate anion. After filtering, the sulfate-bearing filtrate is buffered to a hydrogen ion concentration between pH 2 and pH 3 and zinc precipitated as sulfide. This separates zinc from iron, aluminum, manganese, nickel, cobalt, and chromium, in solution.

For *qualitative detection*, zinc is separated in the same manner as above. The white sulfide is dissolved in nitric acid, some crushed asbestos rolled into a small ball, dipped on a platinum wire into a very dilute solution of cobalt nitrate, $Co(NO_3)_2$, ignited in a gas flame, then dipped into the acid solution being tested for zinc, and again placed into the flame. A green color indicates zinc, a blue color aluminum. A blank test is necessary because a strong cobalt solution may color the asbestos bead.

Solution of the sample. Samples containing not more than 200 mg zinc should be used. Most zinc-bearing substances are easily attacked by hydrochloric or nitric acids followed by addition of 2–3 ml of concentrated sulfuric acid, which is evaporated to fumes and to near dryness to assure the absence of undesirable anions during the hydrogen-sulfide treatment. This also permits an efficient separation of silica if present in the sample. Acid-insoluble zinc-bearing materials are best decomposed by fluxing with sodium carbonate and subsequent acid leaching.

After the separation as sulfide, zinc is usually determined as oxide, ZnO, after ignition at 900°C.

Gravimetric procedure for zinc in silicates

If hydrogen sulfide group metals are present and after the sample has been decomposed by acid attacks and evaporated with sulfuric acid to near-dryness, cool, add 50 ml of 5 N sulfuric acid and digest on a water bath for 10–20 min till all soluble salts have dissolved. Filter the silica off, if present, and wash with about 50 ml of hot 1 N sulfuric acid, collecting the liquid into a 500-ml Erlenmeyer flask. (In the absence of insoluble silica or other precipitates, transfer the original 50 ml of the 5 N sulfuric acid solution from the dish directly into an Erlenmeyer flask by washing with 50 ml of 1 N sulfuric acid). The flask should be equipped with a two-hole rubber stopper for glass tubes. Heat on water bath for about one hour bubbling hydrogen sulfide gas through at intervals, keeping the bottle closed between to create some pressure. Filter the warm solution, after permitting the sulfides to settle, through a fine-porosity filter which has been treated first by passing warm wash solution of the same normality (50 ml 5 N H_2SO_4 + 50 ml 1 N H_2SO_4), saturated with hydrogen sulfide. Wash with the same warm solution, and remove the hydrogen sulfide from the filtrate by boiling (for accurate work, if copper is present, dissolve the sulfides, reprecipitate as above, and combine the filtrates). Neutralize the solution with ammonia till a precipitate just starts forming but dissolves on stirring, evaporate in a marked beaker till the volume is about 100 ml, and return the solution to the Erlenmeyer flask, washing the beaker with 25 ml of distilled water. Add 25 ml of a 20% citric acid solution and neutralize drop by drop with concentrated ammonia, using methyl orange as indicator. Add some acrolein if cobalt is present. Add 25 ml of the formic acid buffer and bring the volume to about 200 ml. Heat on water bath shaking occasionally, and bubble H_2S through for 30 min. Cool with the other glass tube closed while keeping the hydrogen-sulfide pressure on. When the sulfide precipitate settles, decant and filter through a fine- or medium-porosity vitreous silica or porcelain filtering crucible. Wash with a formic acid solution containing 4.6 g/l of the concentrated acid, and saturated with H_2S. Ignite the crucible with the sulfide at 950°C in oxidizing atmosphere by keeping the oven slightly open. Weigh as ZnO.

Gravimetric factor:

$$Zn = ZnO \times 0.80339$$

Reagents. Make the *buffer solution* by diluting 200 ml of concentrated formic acid, HCO_2H (s.g.1.2), to 600 ml. Add 250 g of ammonium sulfate, $(NH_4)_2SO_4$, and, with caution, 30 ml of concentrated ammonia, and bring the volume up to 1 l. Formic acid is dangerously caustic—use rubber gloves.

Titrimetric methods

These are less satisfactory for zinc than the gravimetric procedure described above, but since speed may be a main consideration in routine ore analysis, the *ferrocyanide titration* is given as an example. This method uses potassium ferrocyanide, $K_4Fe(CN)_6$, according to the reaction:

$$3Zn^{2+} + 2[Fe(CN)_6]^{4-} + 2K^+ \leftrightarrows K_2Zn_3[Fe(CN)_6]_2$$

and an internal redox indicator diphenylamine. Previously, an external indicator, uranyl nitrate, for example, has been commonly used.

Procedure

Separate the hydrogen sulfide group as described above, boil, and evaporate the filtrates to 100 ml, add 2–3 drops of a 1% potassium ferricyanide, $K_3Fe(CN)_6$ solution, add 2–3 drops of the diphenylamine indicator. Stir, and titrate with the standard potassium ferrocyanide, $K_4Fe(CN)_6$, solution till the blue color changes to a yellowish-green. Over-titrate slowly by 1–2 ml, and titrate back with a 0.1 *M* standard zinc solution till a sudden color change to purple occurs.

Interferences. Fe interferes in large quantities; nitrates, heavy metals if not removed as prescribed by H_2S, Cd, Mn, and oxidizing substances in general also interfere.

Reagents. 0.05 M ferrocyanide standard solution is prepared by dissolving 21.12 g of the yellow $K_4Fe(CN)_6 \cdot 3H_2O$ in 1 l of water containing 200 mg Na_2CO_3. Permit to stand for 1 week, and standardize with metallic zinc dissolved in 5 N H_2SO_4 and diluted to 0.1 M concentration. Use the same procedure as above. Calculate in terms of mg zinc/1 ml of standard solution.

Estimation of zinc in ores

A *fast* but inaccurate method which permits the *estimation of zinc* in ore samples of complex composition consists of hydrochloric acid attack, followed by nitric acid attack, and evaporation to dryness with 10 ml of concentrated H_2SO_4: Digest the sample with 5 ml of concentrated HCl and 10 g of NH_4Cl on a water bath, dilute to 50 ml and cool. Add 1 ml of liquid bromine and neutralize the mixture with concentrated NH_4OH and bring to boiling with 10 ml of the base in excess. Filter the precipitate, wash immediately a few times with a solution containing 10% NH_4Cl and 10% conc. NH_4OH, return to the same beaker. Dissolve in 5 ml of concentrated HCl, add 1 ml of a 30% H_2O_2 solution, stir to dissolve the manganese, cool, and again add 1 ml of liquid bromine. Neutralize with 10 ml of concentrated NH_4OH in excess, bring to boiling, and again filter hot and wash immediately through a coarse filter paper into the beaker containing the first filtrate. After the addition of 1 g of granular or shot lead, boil for 20–30 min, acidify the solution by concentrated HCl and boil a few minutes more. Add 10 ml of concentrated HCl, dilute to 400–500 ml. Add a few drops of potassium ferricyanide, and titrate with standard $K_4Fe(CN)_6$ solution, using uranyl nitrate as external indicator. When one drop of the titrant gives a brown coloration when added to drops of the indicator placed on a spot test plate, titration is completed.

It is advisable to save a small portion of the titrant and add it to the rest of the solution after the endpoint has been reached, and to proceed cautiously after this in order not to over-titrate. It is important that quantities used and volumes are alike, and it is advantageous to run a similar blank and to subtract the spent volume of the cyanide from the volume spent for the sample before calculating the zinc in the sample.

Reagents. Uranyl nitrate indicator, $UO_2(NO_3)_2 \cdot 6H_2O$, is prepared by dissolving about 5 g of the salt in 100 ml of water.

CHAPTER 8

Classical Volumetric Methods

Analytical procedures which use volumes of solutions of known reacting capacity, rather than weight, are called volumetric or titrimetric methods. These usually consist of the addition of a measured amount of a solution of known strength or "titer" to the sample solution from a pipette. Added *color indicators*, or simply the observation of a color change of the solution when colored ions are present, permit a visual detection of the reaction *endpoint*. The detection of an endpoint may consist of an instrumental measurement of a change of color, or electric property of the solution near the equivalent point when corresponding quantities of the reacting ions have been expended (Fig.15).

NORMALITY AND MOLARITY

When the *normality* of the *standard solution* is known and its volume measured, one can calculate the concentration of the element sought in the sample solution. Solutions of 0.1 N are usually used in routine work. A one-normal solution, written 1 N or simply N, contains one gram-equivalent (1g-equiv.) of the reacting ion in 1 l of solution at room temperature or 20°C. Consequently, accurate volumetric measurements must take the temperature of the "titer" solutions into account. When this is not done, it is assumed that the work is done at room temperature. Of great advantage in such cases is the establishment of a temperature-controlled volumetric room (20° $\pm$ 1°C) where all standard solutions are kept and titrations (which do not involve heating of liquids) conducted. Such a room is shown in Fig.5 which represents a floor plan de-

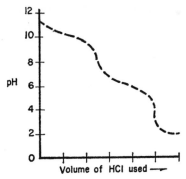

Fig.15. Titration of Na_2CO_3 with HCl showing two endpoints which correspond to dissocation of H_2CO_3 and HCO_3^-, respectively.

signed for accurate geochemical research for (see p.29) the Nevada Mining Analytical Laboratory, Nev. (U.S.A.).

The relationship of gram-equivalents to gram-atoms and of normality to molarity (a n-molar solution, written nM, contains n g-mol./l) is best seen from the following series of equations:

1 g-at. H = 1 g-equiv. H^+ = 1.0 g H
1 g-at. O = 2 g-equiv. O^{2-} = 16.0 g O
$\frac{1}{2}$ g-at. O = 1 g-equiv. O^{2-} = 8.0 g O
$\frac{1}{2}$ g-at. Ca = 1 g-equiv. Ca^{2+} = 20.0 g Ca
$\frac{1}{3}$ g-at. Al = 1 g-equiv. Al^{3+} = 9.0 g Al
$\frac{1}{4}$ g-at. Si = 1 g-equiv. Si^{4+} = 7.0 g Si
1 M^1 Ca solution = 2 g-equiv. Ca^{2+}/l = 40.1 g Ca/l
1 N Ca solution = 1 g-equiv. Ca^{2+}/l = 20.0 g Ca/l
1 M^1 HCl solution = 1 g-equiv. HCl/l = 36.46 g HCl/l
1 N HCl solution = 1 g-equiv. HCl/l = 36, 46 g HCl/l
1 M^1 NaOH solution = 1 g-equiv. NaOH/l = 40.00 g NaOH/l
1 N NaOH solution = 1 g-equiv. NaOH/l = 40.00 g NaOH/l
1 M^1 H_2SO_4 solution = 2 g-equiv. H_2SO_4/l = 98.08 g H_2SO_4/l
1 N H_2SO_4 solution = 1 g-equiv. H_2SO_4/l = 49.04 g H_2SO_4/l

From these equations we can see that the *titer* per ml of a so-

---

[1] Actually "formal" in a strict sense.

lution containing 40.00 g NaOH/l can be expressed as 0.04000 g NaOH, as 0.03646 g HCl, or as 0.04904 g H_2SO_4.

Similarly we can calculate the titer of a 1 N permanganate solution for the *oxidation–reduction reaction* which takes place in acid solution according to the equation:

$$2KMnO_4 + 5H_2C_2O_4 + 3H_2SO_4$$
$$\rightleftarrows 2MnSO_4 + K_2SO_4 + 10CO_2 + 8H_2O$$

or:

$$MnO_4^- + 8H^+ + 5 \text{ electrons} \rightleftarrows Mn^{2+} + 4H_2O$$

or:

$$Mn^{7+} + 5e \rightleftarrows Mn^{2+}$$

thus resulting in a total change of oxidation state of five. One fifth of the formula weight of potassium permanganate corresponds then to one g-equiv. or:

$$KMnO_4/5 = 31.61 \text{ g } KMnO_4$$

Thus, the titer of a 1 N $KMnO_4$ solution for this oxidation reaction will be 0.03161 g $KMnO_4$/ml. If ferrous iron were oxidized to ferric iron by permanganate in acid solution, the corresponding titer for iron should be expressed as 0.05585 g Fe/ml of the N solution, because:

$$Fe^{2+} - e \rightleftarrows Fe^{3+}$$

The titer of a 0.1 N $KMnO_4$ solution for iron would correspondingly be 0.005585 g Fe.

In basic solution, the reduction of manganese in the same reaction proceeds only to the Mn^{4+} state, the equivalent weight thus being $KMnO_4/3$ g. A permanganate solution that is normal in respect to potassium would contain one formula weight, or 158.03 g $KMnO_4$.

Solutions of exactly predetermined normality are difficult to prepare in actual analytical routine. Most of these solutions are also difficult to keep at the initial normality for longer periods of time unless great precautions are exercised. Therefore, it is usually necessary to check the titer of a standard solution with a *primary standard substance* which has to be stable, non-hygroscopic, and have a reasonably high equivalent weight. The most frequently used primary standard substances are the sodium oxalate, $Na_2C_2O_4$, acid potassium phthalate, $KHC_8H_4O_4$, and the ferrous ammonium sulfate hexahydrate, $Fe(NH_4)_2(SO_4)_2 \cdot 6H_2O$. As_2O_3, I_2, KIO_3, NaCl, and metallic Ag are also used for these purposes. All of these substances have to be of ultimate available purity and have to be prepared with great care and kept in desiccators after the removal of moisture. Naturally, in all volumetric work the cleanliness of the glassware and careful calibration and crosscalibration of all byrettes and volumetric flasks are essential.

The other most frequently used permanganate oxidation–reduction reactions in volumetric analysis are:

$$2KMnO_4 + 10KI + 8H_2SO_4 \leftrightarrows 2MnSO_4 + 6K_2SO_4 + 5I_2 + 8H_2O$$

and:

$$2KMnO_4 + 10FeSO_4 + 8H_2SO_4 \leftrightarrows 2MnSO_4 + 5Fe_2(SO_4)_3 + K_2SO_4 + 8H_2O$$

and in *iodimetric titrations*, for example:

$$2Na_2S_2O_3 + I_2 \rightleftarrows 2NaI + Na_2S_4O_6$$

or

$$2S_2O_3^{2-} + I_2 \rightleftarrows 2I^- + S_4O_6^{2-}$$

Accordingly, one gram-equivalent is equal to one formula weight of $Na_2S_2O_3$ in these reactions. All these reactions have a color change near the endpoint in common, and thus do not require an additional indicator.

A commonly used *iodometric* reaction is:

$$2PbCrO_4 + 6KI + 16HCl \rightleftarrows 2PbCl_2 + 2CrCl_3 + 3I_2 + 8H_2O + 6KCl$$

From this reaction we can see that the six-valent chromium is reduced according to:

$$Cr^{6+} + 3e \rightleftarrows Cr^{3+}$$

and that 1 lead is equivalent to 3 iodines, the equivalent for lead thus being $Pb/3 = 69.06$ g. We can also see that 1 ml of a 0.1 N solution of thiosulfate will titrate the liberated iodine equivalent of 0.006906 g Pb. Correspondingly, 100 ml of a 0.1 N thiosulfate solution will titrate 0.6906 g Pb and, if we wanted to standardize such solution, we could weigh pure lead foil, dissolve it, and calculate the normality on the basis of the consumed ml of the titer solution.

In *chromate titrations* of iron the main reaction can be written:
$$K_2Cr_2O_7 + 6FeCl_2 + 14HCl \rightleftarrows 2CrCl_3 + 6FeCl_3 + 2KCl + 7H_2O$$
where the chromium oxidation state changes:
$$Cr^{6+} + 3e \rightleftarrows Cr^{3+}$$
and the gram-equivalent weight of $K_2Cr_2O_7$ is thus equal to $K_2Cr_2O_7/6$ g.

Procedure for total iron

Put 10 ml of concentrated phosphoric acid and 10–20 ml of water into a 600-ml beaker. Pass the iron-bearing solution (p.75) through a silver reductor (p.80) into this beaker, keeping the delivery tip under liquid to prevent oxidation by air. Wash the reductor with five 10-ml portions of a 1 N HCl solution, collecting the liquid in the same beaker. Close the stop cock and lift the delivery tube from the solution, pass two to three more 10-ml portions of the HCl solution, and rinse the tip that was immersed in the liquid. Add 2–4 drops of a 0.2% solution of diphenylamine, in concentrated H_2SO_4, stir, and titrate with a 0.1 N $K_2Cr_2O_7$ solution to a violetblue color that persists for about 30 sec.

Note that acid stronger than 1 N causes the formation of hydrogen which may form a gas-lock in the reductor, thus obstructing the flow of the liquid. Hydrogen peroxide, if not removed by boiling, passes through the reductor and oxidizes one part of the iron. Manganese, titanium, chromium, and vanadium do not interfere.

Copper interferes and must be separated by hydrogen sulfide or diethyldithiocarbamate precipitation before the ammonium hydroxide precipitation. Most rocks do not contain, however, more than 0.01% Cu. Copper catalyzes the oxidation of ferrous iron in the presence of air oxygen.

For the *preparation* of a 0.1 N $K_2Cr_2O_7$ solution, dry the highest purity salt at 150°C for 2 h, cool, weigh 4.9032 g, dissolve in distilled water in a 1-l volumetric flask and fill to the mark. The solution can be further standardized against electrolytic iron wire.

Acid-base titrations

Acid-base titrations are essential in volumetric methods which are used in order to determine the exact hydrogen or hydroxyl ion concentrations in solutions. Important information can be obtained by these methods in the analysis of carbonate and bicarbonate mixtures, ammonium salts and other buffers, phosphates, and in the study of different indicators.

Hydrochloric acid, sodium hydroxide and anhydrous sodium carbonate (HCl, NaOH, Na_2CO_3) are the most frequently used standard substances in *acidimetry* or *alkalimetry*, as this branch of volumetric analysis is often called. The method consists of neutralization titration with measured volumes of known strength. The endpoints are determined by the sudden change of color of indicators. This color change can occur at selected pH values, depending on the indicator. It is usually due to sharp changes of the hydrogen ion concentration near this point.

Water dissociates into hydronium and hydroxyl ions according to the reaction:

$$2H_2O \rightleftharpoons H_3O^+ + OH^-$$

conventionally represented as:

$$H_2O \rightleftharpoons H^+ + OH^-$$

Changes *in hydrogen ion concentration*, especially near the neutral point, have profound effects on chemical reactions. When the concentration of hydrogen ions exceeds that of the hydroxyl ions, we call the solution acid. In basic solutions, the reverse is the case. A neutral solution has an equal concentration of hydrogen and hydroxyl ions. We can see from this that water participates directly in neutralization and volumetric titrations in aqueous solutions in general. The effect of water on dissolved salts of weak acids and strong bases is called *hydrolysis*. This results in basic solutions even though equivalent amount of both acid and base may be present. Solutions of salts of weak bases and strong acids give acid reactions due to the production of hydrated protons and this process is called *hydration*.

According to the law of mass action, the *ion product constant* for water can be expressed by equation:

$$K_w = [H^+][OH^-] \qquad (7)$$

This dissociation constant K_w depends on the temperature of the solution. At 25°C there are 10^{-7} hydrogen ions in pure electrolyte-free water and there is the same amount of hydroxyl ions. By taking a negative logarithm of both sides of the preceeding equation, we can write:

$$-\log K_w = -\log [H^+] -\log [OH^-] \qquad (8)$$

or

$$-\log K_w = pH + pOH \qquad (9)$$

if p is defined as the negative logarithm of the molar concentration of the respective ions in the solution. Therefore:

$$-\log K_w = 7 + 7 = 14 \qquad (10)$$

We can now express the hydrogen ion concentration, conventionally written as pH, as a fraction of the number 14 where diminishing values below seven ($<$pH 7) would indicate an increasingly acid solution and values above seven ($>$pH 7) correspondingly a basic solution.

To *calculate a pH value* of a solution with a hydrogen ion concentration of $5 \cdot 10^{-6}$, for example, we write:

$$pH = -\log 5 \cdot 10^{-6} = -\log 5 -\log 10^{-6}$$
$$pH = -0.7 -(-6) = 5.3$$

or:

$$pOH = 14-5.3 = 8.7$$

We can also see that the p-function of an ion of greater concentration than 1 *M* will be a negative number.

Standardization of acid or base solutions can be performed using appropriate indicators. The effect of atmospheric carbon dioxide which is always present must be considered in accurate work and the choice of proper indicators also is influenced by this factor. For example, water in equilibrium with normal atmosphere having about 0.03% CO_2, has a pH value of about 5.7, and only carefully deionized water that has been kept isolated from the atmosphere has the theoretical value of pH 7.

Titrating strong acids with strong bases is the most straightforward procedure because only the reaction:

$$H^+ + OH^- \rightleftarrows H_2O$$

need be considered. For example, since we may assume complete dissociation of the hydrochloric acid in water, we can write:

$$HCl \rightleftarrows H^+ + Cl^-$$

The hydrogen ion concentration will then depend on the molarity of the solution which is equal to the normality in this case:

$$[H^+] = N \tag{11}$$

or

$$pH = -\log N \tag{12}$$

so that if we have a 0.1 N HCl solution, we can calculate its pH:

$$pH = -\log 1 \cdot 10^{-1} = -0 - (-1) = 1$$

Standardization with sodium carbonate

Weigh 250.0 mg of standard grade Na_2CO_3 which has been kept in dessicator after drying for about 2 h at 150°C. Dissolve in 50 ml of distilled water and add 2–3 drops of bromcresol green indicator which undergoes a color transition between pH 4 and pH 5. Titrate with hydrochloric acid till first change of color from bluish to green, boil the solution for 2 min to remove the released CO_2, cool the now blue solution quickly by placing in cold water, and complete the titration of the remaining bicarbonate ion till a permanent yellowish-green color. For accurate work, determine the indicator blank by titrating as many drops of it in a solution of the same volume and containing about the same amount of NaCl.

Note that sodium carbonate solutions have two endpoints, the first between pH 8 and pH 9, and the second more sharp endpoint at about pH 4. This latter endpoint is always used. The corresponding reactions are:

$$H_2CO_3 \rightleftarrows H^+ + HCO_3^-$$
$$HCO_3^- \rightleftarrows H^+ + CO_3^{2-}$$

The titration curve for sodium carbonate is given in Fig.15. Methyl orange or methyl red can also be used as indicators in this titration. One can also over-titrate till yellow color of bromcresol green, boil to expel CO_2 and thus restore the equilibrium, and titrate back with a standardized base. If the normality of the base is different from the normality of the acid used in the titration, the expended volume of the base has to be multiplied by the ratio of its normality to the normality of the acid used, before correcting for the back titrated volume by subtraction.

Preparation of 0.1 N sodium hydroxide solution

Sodium hydroxide solutions absorb carbon dioxide from the

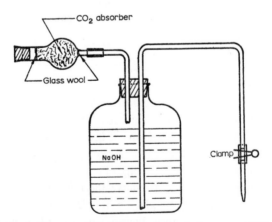

Fig.16. Bottle for storage of NaOH titrimetric solutions to avoid the effect of atmospheric CO_2.

TABLE IV

DILUTION OF CONCENTRATED ACIDS AND AMMONIA TO OBTAIN
APPROXIMATELY 5 N SOLUTIONS FOR USE AS SHELF REAGENTS

	HCl	H_2SO_4	HNO_3	CH_3COOH glacial	NH_4OH
Specific gravity	1.2	1.8	1.4	1.1	0.9
Percent reagent	36.0	96.0	69.5	99.5	58.6
Normality	~12	~36	~16	~18	~15
Milliliters diluted to 1,000 ml to obtain ~5 N solution	430	140	320	290	330

atmosphere. This complicates their preparation and storage. Plastic bottles for NaOH solutions have to be equipped with a rubber stopper with two tubes attached. One of these is partially filled with a CO_2–absorbing substance, Ascarite for example, and serves as air intake when the solution is used (see Fig.16). 1–2-liter flasks are convenient to use for storage.

Procedure

Promptly weigh about 8 g NaOH in a 125-ml Erlenmeyer flask, add 20 ml of icecold, freshly distilled, and boiled water, swirl to dissolve, stopper, and cool quickly on ice. Filter immediately through a clean Gooch crucible. Use about 10 ml of filtered solution per liter of cold CO_2-free water, placing the solution immediately into the storage bottle. Standardization of this solution is best performed by acid potassium phthalate, $KHC_8H_4O_4$.

Standardization of sodium hydroxide solution

Dry standard grade $KHC_8H_4O_4$ for 2 h at 105°C, and weigh 500–1000 mg of the reagent. Dissolve in 40 ml of freshly-boiled cold water, add 2–3 drops of phenolphthalein, which has transition point at about 8–10 pH from colorless to red. Titrate till pink color persists for 30 sec. The one "acid" hydrogen in the phthalate is replaced by sodium in this reaction, the formula weight thus represents 1 equivalent of NaOH.

Phenolphthalein indicator solution is prepared by dissolving 500 mg of the reagent in 1 l of an 80% ethanol solution.

Preparation of approximately 0.1 N hydrochloric acid

Dissolve 8.1 ml of analytical grade concentrated HCl in cold freshly boiled water and dilute to 1 l. Mix well and titrate with standard NaOH if the acid is to be used in titration.

Preparation of *approximately 5 N acids* and *ammonia* to be used as shelf reagents is done by diluting the volumes of concentrated reagents given in Table IV to 1 l. Most procedures given in this book are based on the use of the 5 N solutions as such or further diluted.

The silver reductor

The silver reductor consists of a column 30–50 cm long and about 2 cm in diameter filled to a height of about 10–20 cm with freshly precipitated silver of about 20-mesh size from which the fines have been removed by screening (Fig.17). The precipitation of silver is performed on a zinc rod overnight, from 1 l of a 3% silver nitrate solution to which 1 ml of nitric acid and 30 ml of glacial acetic acid are added. The precipitated washed silver is loaded over a glass wool mat into the reductor to pass about 40–50 ml of solution per min and kept covered at all times with a 1 N solution of HCl. If commercial electrolytically precipitated silver of about 20–mesh grain size is available, the precipitation of silver may be omitted. The used portion of the silver column becomes black from reduced precipitated silver chloride. When about two thirds of the column are darkened, the silver may be replaced, or regenerated by ammonium hydroxide treatment and a short bath in a 5 N nitric acid solution, followed by washing if commercial shot-silver was used.

The Jones reductor

The *Jones reductor* column may be longer than the silver column (which is expensive). It is otherwise identical in construction with

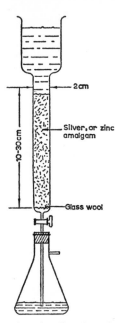

2cm

Silver, or zinc
amalgam

10-30cm

Glass wool

Fig.17. The silver reductor. When the silver is replaced by amalgamated zinc,
it is called the Jones reductor.

Fig.17. The zinc metal used is about 20 mesh in grain size and
should be iron-free. The zinc is amalgamated with mercury to
inhibit the formation of hydrogen gas which easily blocks the flow
of liquid through the column if formed. The *amalgamation* is
performed by covering the zinc with a 1 *N* HCl solution and stirring for 30–60 sec, decanting the acid, and covering the wet zinc
with a solution containing 80 g/l of mercuric nitrate, $Hg(NO_3)_2$, or
70 g/l of mercuric chloride, $HgCl_2$. Stir vigorously for 3–5 min,
decant and wash three to five times with water. Load the amalgam
into a water-filled reductor to a height of about 30 cm, pass 300–
500 ml water, and keep the amalgam always under water. One normal sulfuric acid solution (1 *N* H_2SO_4) is preferred but the sample
solution may contain from 1 to 10% of H_2SO_4. One to five normal
(1–5 *N*) hydrochloric acid solutions may also be used. Nitrates
should be absent because they oxidize iron.

CHAPTER 9

Emission Flame Spectrophotometry

INTRODUCTION

Flame excitation for emission photometric analysis was introduced by Lundegård in 1929. Spectrochemical methods in general
are based on the excitation of electromagnetic radiation from atoms
by the heat of a flame or electric arc or spark. This radiation can
be separated into an array of spectral lines or wavelengths typical
for each element. The intensity of a particular spectral line is
related to the concentration of the element in the sample. In flame
spectrometry excitation is produced by subjecting the sample solution to the heat of a flame through atomization or spraying.
Different flames of varying temperature are used. The most common are air-acetylene, air-gas, oxygen-acetylene, air or oxygen-
hydrogen flames. The hydrogen air flame gives very low background. This type of excitation produces relatively few spectral
lines due to the relatively low thermal energy of the flame. For
example, the flame spectrum of sodium, or the yellow light that we
see when sodium compounds are burned, consists of radiation at
wavelengths of 5895.9, 5890.0, 3303.0, and 3302.0 Å, respectively.
The first two of these lines give the yellow color to the flame.
Elements whose atomic spectra are thus excited are Li, Na, K, Rb,
Cs, Cu, Mg, Ca, Sr, Ba, Zn, Cd, Hg, Sc, Y, lanthanides (rare
earths), Ga, In, Tl, Pb, Cr, Mn, Fe, Co, Ni, Ru, Rh, and Pd. The
molecular spectrum of boron is also used. Qualitative and quantitative determination of these elements can be performed by flame
spectrophotometry.

INSTRUMENTATION

Instruments employed are either spectrographs, which use prisms or gratings to separate the light into its compounds and record these in the form of spectral lines on photographic film, or spectrophotometers, which record the radiation directly using light-sensitive electric devices. In these, prisms or gratings or simply filters select and separate the desired wavelengths. In flame work, spectrophotometers are presently more common than spectrographs, and we speak of *flame photometry* as the technique. Fig.18 shows how flame spectrophotometry works.

In quantitative analysis the blackness or darkening degree produced on film or the optical density of the slit image in the spectrograph is a measure of intensity of the radiation at that wavelength. We measure this density with microphotometers or densitometers. It can be converted into relative intensity and plotted on a calibration curve versus known concentrations of the element in the sample. In this manner the so-called working curves are produced (Fig.24).

When relative intensities of specific radiation of unknown samples are plotted on these curves one can estimate the concentration of the respective element in the sample, assuming that the dilution factor is known.

Selecting proper instruments for flame-photometric work depends on the complexity of the sample and on the sample preparation method. When only a few elements are to be determined and no interfering elements are present in the original sample, relatively unsophisticated equipment may be used, employing

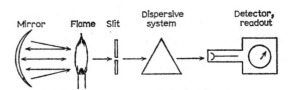

Fig.18. Schematic presentation of the emission flame analysis equipment and the principle of the method.

simply filters. This is the case when only alkalies or alkaline earths are to be determined.

When complex mixtures are to be analyzed by flame, more sophisticated equipment with better dispersion, selectivity, and sensitivity has to be used. Most elements can be determined satisfactorily with the modern prism or grating flame spectrophotometers equipped with photomultiplier tubes. The emission flame spectrophotometry of complex samples suffers greatly from *interelement interferences* in the spectrum as well as from *enhancement* and *suppression* due to the presence of other compounds and ions. This means almost unavoidable chemical separation preceding the spraying and burning of the solution, if accurate work is to be done. This again means in most instances that the composition of the solution to be analyzed will be simple enough to give at least satisfactory results using good relatively inexpensive equipment. Fig.19 shows a logical cross-section of the Beckman Model DU Spectrophotometer which is one of the most frequently used machines in this field.

An analyst working in a small laboratory with straightforward routine procedures will often discover that sophisticated equipment capable of a whole spectrum of application is often costlier by an order of magnitude, yet is more difficult to operate and keep in order, and is actually seldom used to its fullest extent. Nevertheless, the tendency has been to stretch the application of a method beyond its natural limits. Whereas this is commendable in

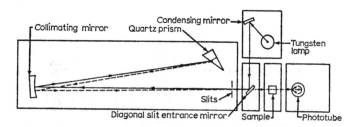

Fig.19. Diagram of the Beckman Model DU optical system. To understand this schematic, follow the light path starting with the tungsten lamp. (Courtesy of Beckman Instruments, Inc.)

terms of exploring new possibilities and areas of marginal application in order to delineate a method's usefullness or "universal"
nature, an experienced analyst tries to avoid working under un-
favorable conditions, and switches whenever possible to another
more suitable and overlapping method. For example, Na, K, and
Ca can be determined very well by emission flame spectrophotometry, Mg can also be determined, but due to weak emission
characteristics and a strong self-absorption effect at high concentrations, only fair results can be obtained using this method for
magnesium. Thus, a switch to atomic-absorption flame photometry is fully warranted when high accuracy is desired and frequent
magnesium determinations have to be performed.

The contents and structure of this book are based on this type of
consideration. To further expand this line of thought, we may add
that in a small laboratory not equipped with either emission or
absorption apparatus, but mainly concerned with routine determinations of Na, K, Ca, and Mg, it would seem advisable to first acquire basic atomic-absorption apparatus rather than emission apparatus because in addition to Mg, also Na, K, and Ca can be
determined satisfactorily by the absorption method.

The preparation of *working curves* as well as the *procedure of
additions* are identical with the methods described under atomic absorption (Chapter 10).

Detection limits of elements frequently determined by flame spectrometry and atomic absorption are given in Table V, p.186. Most
spectrophotometric equipment is provided with lists of detection
limits of elements. The sensitivity depends on specific analytical
conditions, such as the type and temperature of the flame, the
burner, the burner slit and its width, the electronic multiplication,
the type of attachments if used, and the composition of the solvent.
It also depends on gas flow rates, pressure and the solution volume
sprayed into the flame per unit time. In addition, the emission
intensity depends on the spectral line used.

In general, detection limits for alkali metals and the alkaline
earths are in the vicinity of 1 p.p.m. in solution or better. Beckman
Instruments Inc. defines the *detection limit* of an element as the

TABLE V

APPROXIMATE DETECTION LIMITS IN ABSORPTION
AND EMISSION SPECTROSCOPY

Element	Absorption (mg/ml in sol.)	Emission (mg/ml in sol.)
Al	0.1	0.2
Ba	0.1	0.03
B	20.0	0.1
Ca	0.005	0.005
Cr	0.01	0.01
Cu	0.01	0.03
Fe	0.02	0.1
K	0.002	0.003
Mg	0.003	0.04
Mn	0.01	0.01
Na	0.005	0.0001
Ni	0.02	0.6
Pb	0.05	1.0
Si	0.2	10.0
Sr	0.02	0.004
Ti	0.2	0.5
Zn	0.005	20.0

concentration limit that produces a response 1% above the flame background or blank reading. In practical work, the lower detection limits are reached when a concentration of an element causes 1% deflection on the meter scale used, and when this deflection reaches and equals the mean-square-root of the fluctuation of the background. This means a possible error of $\pm\,100\%$ at that concentration. Detection limits given in compiled lists are based on optimum conditions and do not take in account the possible spectral interferences, enhancement or absorption effects in an actual sample. Detection limits depend on sensitivity characteristics of the method, the equipment, and the conditions of the actual experiment. Thus, detection limits for the same element can vary significantly.

Generally, the values determined as detection limits can be used to estimate the *error* of the method because one could not expect a better precision than the maximum detection limit. Under these conditions, precision and accuracy have the same order of magnitude. Thus, we can say that working with a detection limit of 1 p.p.m., for example, a concentration of an element equivalent to 100 p.p.m. can be expected to be determined with a ± 1 p.p.m. accuracy under optimum conditions and we would report 100 ± 1 p.p.m. or $\pm 1\%$.

Tables of detection limits can be used to estimate whether the desired accuracy can be achieved. For example, if we had a sample containing about 100 p.p.m. of magnesium and wanted to determine it by flame emission with a 1% accuracy, knowing that the detection limit of the method is 10 p.p.m., we could immediately see that this accuracy could not be achieved, because 10 p.p.m. is 10% of 100 p.p.m., and we could not expect a better accuracy.

If, on the other hand, we had a solid sample containing 1000 p.p.m. of calcium, which we dissolved in 10 ml of acid, and wanted to know whether we could determine calcium in this 100-p.p.m. solution with an accuracy of 1 p.p.m. or 1%, knowing the detection limit of our procedure to be 0.5 p.p.m., we could assume that this would be possible because a 0.5-p.p.m. detection limit permits us to expect an accuracy of ± 0.5 p.p.m., or $\pm 0.5\%$. This, of course, assuming that the sensitivity of the method and the instrument permits us to read 1 part in 200, and that the fluctuations of the background do not constitute a larger deflection than $\pm 0.5\%$. In practical work, it is advisable to assume a $\pm 1\%$ deflection for the background, which assures the proper reading of most meters. The importance of proper dilution of the sample for maximum accuracy can be seen from this example.

In order to simplify, the concentrations above were given in p.p.m. In solutions concentrations are conventionally given in mg/ml. Both notations can be considered identical as long as the specific gravities of solvents remain similar.

Preparation of sample

In geochemical work, when silica is present, the HF + HClO$_4$ attack and double evaporation to dryness with 5-ml portions of concentrated HCl is the most common first step of sample preparation. For cement analysis, an attack with diluted HCl is often sufficient (see atomic absorption p.201). The size of the powdered sample is usually 100–500 mg and the final dilution will depend on the concentration of the element to be determined. The final concentration of hydrochloric acid in a sample solution prepared after the removal of interfering anions, is kept between 1 and 3 N and it should be the same in all sample solutions compared (see below). Generally, a neutral or barely acid solution is preferred for spraying in order to minimize the possible damage to burners. Ca, Mg, and Ba can be kept in neutral solution by addition of equivalent amounts of NH_4Cl to sample and standard.

Removal of interfering anions

In addition to serious spectral interference from numerous cations, such common anions as sulfate and phosphate seriously interfere in flame analysis by depressing the intensities of magnesium and calcium. These acids, when present in the sample, also may cause partial or complete precipitation of several cations, especially in nearly neutral solutions, and are, therefore, best removed by passing the sample solution through an *anion exchange column:*

To *prepare the column* (Fig.20), use strongly basic Dowex 2, Duolite A-102 D, or Amberlite IRA-410 anion exchange resins. Regenerate the expanded resin, which should be in the OH-form, by agitating about 50 ml of the resin with about 200 ml of a HCl solution prepared by diluting 100 ml of a 5 N HCl solution to 200 ml. Fill the column with this slurry to a desired height and wash three to four times with 30-ml portions of freshly boiled cold water.

Remove the *sulfate* and *phosphate* ions by passing the clear sample solution through the column drop-by-drop, rinse the column three to four times with 10-ml portions of cold boiled water, and dilute the sample solution that now contains dominantly chloride anions to the desired volume. Spray the chloride solution into the flame and bracket the unknown with similarly treated standard samples.

The column may again be *regenerated* by passing 3 N HCl so-

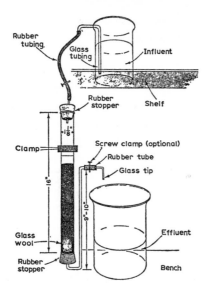

Fig.20. Versatile apparatus for ion exchange which can easily be adapted to different types of chromatographic separations. For example, in the case of removal of interfering anions prior to flame photometric analysis, one replaces the upper beaker by a separatory funnel. (Drawing courtesy of Duolite Resins-Diamond Alkali Co., Western Division.)

lution and rinsing with boiled CO_2-free water.

In emission flame spectrophotometry as in atomic absorption, one must compare samples dissolved in identical acids of the same concentration. Avoiding the burning of samples containing unknown mixtures of acids or organic substances in varying concentrations spares the analyst from disappointing surprises. When using commercially available standard solutions, it is important, therefore, to know the acid used and the solvent. Buffered solutions and complexing organic substances, often added to assure the homogeneity of the solution, have to be reproduced in the preparation of the sample solution as closely as is practically feasible.

Numerous sample preparation methods are designed so as to prevent the interfering anions and cations from contaminating the sample solution. For example, if only alkalies and alkaline earths are to be determined by flame photometry, Shibata's (p.58) ion exchange method of separation will provide a fast and satisfactory separation and an ideal solution of Na, K, Ca, and Mg ions.

Determination of sodium, potassium, calcium (and magnesium)

Excitation of sodium, by flame from the ground state to the lowest excited state occurs in the conventional gas-air flame. The transition of excited atoms to the ground state in this type of flame is the only process recorded. Work with sodium may be, therefore, simplified by controlling the flame temperature to avoid the excitation of other atoms present in the solution. The spectral resonance lines listed for different elements in Table VI can also be used for emission spectrophotometric purposes. These are usually the most sensitive lines, and when higher elemental concentrations are used, one must switch to less sensitive lines listed in equipment manuals and most textbooks. In the case of alkali metals and the alkaline earths, one has, however, to take in account considerable *self-absorption* at higher concentrations and *ionization* which takes place at low concentrations. Both of these phenomena necessitate some experimentation with flame temperatures by adjustment of the fuel/air ratio in order to find the best analytical conditions. Therefore, sodium is best determined by selecting the lowest possible flame temperatures that will efficiently excite this element. Since the elements in question all have relatively low ionization potentials varying between 3 and 6 eV, the spectral emission lines are mutually very sensitive to the varying concentrations of these elements. This often makes it necessary to further separate chemically the alkalies from the alkaline earths before an accurate flame emission spectrographic determination can be performed. If, however, the relative concentrations of the matrix vary within narrow limits, which is often the case when similar compounds are analyzed in geochemical research or for industrial purposes, one may proceed comparing line intensities of elements in mixtures dissolved without time consuming chemical separations, if similar standards are available.

Selection of optimum conditions in relation to the above mentioned possible effects and interferences is of basic importance in the flame spectrographic analysis. The use of the linear parts of the calibration curves is recommended. However, by closely bracketing the sample between similar standards, even the curved segments of the curves may be utilized in less demanding work.

Standard solutions listed in Table VI may be used, with some exceptions where the detection limits make the use of low concentrations not feasible as is the case of *magnesium* for example. In addition to pronounced self-absorption, the magnesium resonance line at 2,852 Å lies within an OH-band. This coincidence makes the determination of magnesium by flame spectrophotometry much less satisfactory than that of the other elements in group II. One preferably records the whole spectral region between 2,830 and 2,870 Å in order to determine magnesium. Due to the high background, the detection limits for magnesium by this method are relatively high, varying usually between 10 and 50 p.p.m. The determination of magnesium by atomic absorption is therefore preferred.

Beer's law

In emission flame spectrophotometry, as well as in atomic absorption and colorimetry an analyst assumes that the absorbance of a flame or solution is directly proportional to the concentration of the absorbing species in that medium. In most cases, this theoretical relationship, called *Beer's law*, is true when the measurement is performed at the same distance from the source and through the same thickness of the medium for both the sample and standard. This law is expressed by eq.13:

$$A = \log \frac{I_0}{I} \tag{13}$$

where A is *absorbance*, I_0 is intensity of the incident beam, and I is the intensity of transmitted radiation. *Transmittance* equals the fraction of incident radiation that passes the sample medium:

$$T = \frac{I}{I_0} \tag{14}$$

TABLE VI

Element	Stock sol. (mg/ml metal, as chloride)	Standard sol. (mg/ml metal, as chloride)	Line (Å)	Slit (mm)	Oxidiz (fuel C₂
Al, metal	1,000	0,10, 50, 100, 150, 200	3093	0.3	N_2O
Ba, BaCl$_2$	1,000	0, 25, 50, 75, 100, 150, 200	5536	0.3	N_2O, or air
B, B$_2$O$_3$	5,000, Na borate	50, 100, 200, 400, 800, 1,000	2497–8	1.0	N_2O
Ca, CaCO$_3$	500	0, 2, 5, 8, 10, 15, 20, in ~1% La solution in 5% HCl	4227	1.0	air, c N_2O
Cu, metal	1,000, nitrate	0, 5, 10, 15, 20	3248	1.0	air
Fe, metal	1,000	0, 5, 10, 15, 20	2483	0.3	air
K, KCl	500	0, 2, 5, 8, 10	7665	1.0	air
Mg, metal	500	0, 0.5, 1.0, 1.5, 2.0	2852	3.0	air
Mg, metal	1,000	0, 2, 5, 10, 15, 20, 25	2852	0.5	N_2O
Mn, metal	500	0, 2, 5, 10, 15, 20	2795 (2801)	1.0	air, o N_2O
Na, NaCl	500	0, 0.3, 1, 2, 3	5890–5896	0.3	air
Ni, metal	500	0, 2, 5, 10, 15, 20, 25	2320	0.3	air
Pb, metal	1,000, nitrate	0, 5, 10, 20, 30, 40	2170, 2833	1.0	air, o N_2O
Si, Na salt	5,000, Na$_2$SiO$_3$	0, 50, 100, 200, 300, 400, 500	2516	0.3	N_2O
Sr, SrCl$_2$	1,000	0, 5, 10, 15, 20	4607	1,0	N_2O ◆ air
Ti, TiO$_2$	5,000	0, 5, 10, 25, 50, 100, 200, 400	3643	0.3	N_2O
Zn, metal	500, nitrate	0, 0.5, 1.0, 2, 3, 4, 5	2139	1.0	air, o N_2O
Cr, metal	1,000	0, 2, 5, 10, 15, 20	3579	0.3	N_2O

Flame	Interference	Sensitivity (mg/ml at 1% abs.)	Remarks
uminescent, reducing	add 1,000 p.p.m. Na to sample and standard	>1	add 50–100 mg Hg when dissolving in HCl
uminescent, reducing	add 1,000 p.p.m. Na to sample and standard; if air is used: Al, P	>5	use N_2O
uminescent, reducing	—	~50	—
uminescent, reducing	use N_2O, or add 1,000 p.p.m. Na to sample and standard if air is used	~0.1 >0.1, N_2O	dissolve in dil. HCl
n-luminescent, oxidizing	—	$\lesssim 0.1$	—
n-luminescent, oxidizing	—	0.1; 1.0, 3719.9 Å	—
n-luminescent, oxidizing	enhanced by Na, Li, Cs	>0.1	Osram lamp
uminescent, reducing	Al, P, Si, SO_4^{2-}	>0.1	—
uminescent, reducing	minimal	>0.1	add 1% Sr or La, if Al interferes
n-luminescent, oxidizing	(Si)	~0.1	2nd line, reduced sensitivity
n-luminescent, oxidizing	minimal	>0.05	Osram lamp
n-luminescent, oxidizing	—	>0.1	3415 Å line less sensitive
n-luminescent, oxidizing	—	~0.5	2614 Å line less sensitive
uminescent, reducing	—	>3.0	more work needed
uminescent, reducing	Si, Al, P if air is used	>0.2	use N_2O, or add 1,000 p.p.m. Na and 1% La, see Ca
uminescent, reducing	—	~2.0	—
n-luminescent, oxidizing	(Si)	~0.05	—
uminiscent, reducing	—	~0,2	4254 Å line less sensitive

CHAPTER 10

Atomic Absorption Spectroscopy

INTRODUCTION

Atomic absorption spectroscopy, introduced by WALSH (1958) in 1955 as an instrumental analytical technique, complements and has certain advantages over the emission flame photometric methods first applied by Lundegårdh in 1929. Despite the late start, this technique has found wide and rapid acceptance by analysts in varying fields. Unlike the emission methods which are based on the excitation of atomic species by the heat of a flame and the measurement of emitted radiation as characteristic spectra, atomic absorption consists of the measurement of absorption caused by chemically unbound atoms in a flame. Whereas not more than 4% of atoms of an element are raised to a high-energy level in conventional burners, the rest of the atoms, in an ideal case, are merely stripped from their chemical bonds and remain at the ground state. These atoms absorb radiation in a few extremely narrow bands of about 10^{-4} Å in width. These resonance lines occur at a different wavelength for each element. The actual width of these bands is increased due to Doppler broadening.

It is evident that atomic absorption should theoretically be more sensitive because the amount of atoms causing the effect may be orders of magnitude larger than in emission. The relative predominance of atoms in the minimum-energy state over the high-energy state atoms in the flame also assures higher precision and better stability, since variations in flame temperature produce negligible changes in the number of neutral atoms.

Although there is only a minimal or practically no spectral interference by other elements caused by emission in the wavelength region of the element that is being determined, atomic absorption as an analytical method does have to consider inter-element effects due to the physico-chemical nature of the sample. For example, elements forming chemical compounds or refractory oxides such as aluminum, tungsten, vanadium, and titanium, or temperature-stable compounds such as phosphates, sulfates and silicates, may have depressive effects on absorption. Also, ionization of elements with low ionization potential (<6 eV) may remove appreciable amounts of ground state atoms from the flame, causing a decrease of the absorption effect.

These chemical interferences are usually overcome by either: (1) the chemical removal of the offending constituents; (2) the addition of an excess of the interfering cation or anion to the standard and the unknown sample, or the use of a releasing agent such as lanthanum; (3) the use of a complexing agent, usually EDTA, in order to prevent the formation of refractory compounds; or (4) the use of a different gas mixture for the flame for dissociation purposes, for example the nitrous oxide-acetylene mixture.

Enhancement of absorption is known to occur when an easily ionizable element is present or introduced. This shifts the equilibrium between the ionized and ground state atoms in the direction of the neutral atoms. Organic solvents also enhance absorbance about four-fold by increasing vaporization of the solution. They also facilitate the extraction of the metals from the solution.

INSTRUMENTATION AND METHODS

The basic equipment for absorption spectroscopy consists of a *hollow cathode lamp*, an *atomizing system* and a *burner*, a *monochromator*, and a *photodetector* arranged according to Fig.21.

In this system, the hollow cathode lamp is chosen to emit light in a narrow band width at the same wavelength as that of the metal being analyzed. This source radiation, called "background" radiation by some authors, is absorbed in the flame of the burner by the atoms of this metal. The atoms, which are excited by this radi-

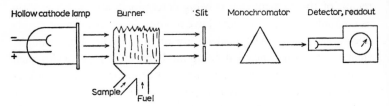

Fig.21. Schematic presentation of atomic absorption equipment and the principle of the procedure.

ation, return rapidly to the ground state emitting *fluorescent radiation* in all directions and at a wavelength corresponding to the source. Thus absorption occurs in this case at the same wavelength as emission. The narrow bands where absorption and emission coincide are called *resonance lines*. Fig.22 shows how atomic ab-

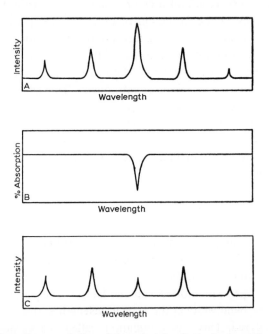

Fig.22. Schematic presentation of the principle of atomic absorption analysis demonstrated by the comparison of emission and absorption spectra of an element. A. Emission spectrum. B. Absorption spectrum. C. Resultant spectrum.

sorption works. The absorption of the source radiation is proportional to the amount of atoms present in the ground state, thus obeying Beer's law. The concentration of the metal in the sample is proportional to absorbance, A, which is proportional to the logarithm of the reciprocal of sample transmittance, T:

$$A = \log\left(\frac{1}{T}\right) = -\log T \qquad (15)$$

The light from the hollow cathode lamp is partially absorbed passing through the flame, and the absorbance or transmittance of this radiation is measured in a photodetector readout system. To select the proper wavelength corresponding to the element that is being analyzed, a monochromator, a grating system, or simply a filter, may be used. These devices reduce the amount of background radiation, thus increasing the proportion of the absorbed light. To further decrease the contribution of the light emitted in the same wavelength region by the flame, modulators and choppers are used which can be adjusted or electrically coupled to the photomultiplier, making it respond only to the desired light component. Different devices which disperse the sample into the flame have been developed. Among these, the most frequently used are the so-called "laminar flow", the "total consumption", and the "turbulent" burners. Recently, nitrous oxide (N_2O)-acetylene burners, which are flashback-free, have appeared on the market, designed by the Fisher Scientific Co.

Atomic absorption apparatus on the market is of widely varying degrees in sophistication and should be selected according to specific needs. Presently, hollow cathode lamps are commercially available for almost all elements.These sealed tubes usually consist of a tungsten anode and a cylindrical hollow cathode lined or made of the element which is to be determined (Fig.23). For volatile elements, Na, K, Rb, Cs, etc., a different type of arc can also be used as a light source. These emission lamps are manufactured by Osram. New high-intensity hollow cathode lamps and multi-element lamps are now commercially available.

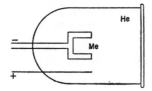

Fig.23. Schematic of a hollow cathode lamp. The cup-like cathode is usually made of the metal which is being analyzed.

Methods

Elements Al, Ba, Ca, Cd, Co, Cr, Cu, Fe, K, Mg, Na, Ni, Pb, Rb, Si, Sr, Ti, and Zn have been determined by atomic absorption with sensitivities of 1 p.p.m. or better. Table V, p.186 gives approximate detection limits in absorption and emission spectroscopy, and Table VI, p.192–193, the approximate conditions for determination of major elements listed above, using the Perkin Elmer Model 303 instrument, respective hollow cathode sources, acetylene as fuel, and air or nitrous oxide as oxidizers. The analyst must

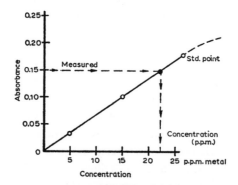

Fig.24. A working curve for the atomic absorption analysis. Absorbance of standard solutions of known concentration, in flame, is measured and plotted against the known concentration expressed in p.p.m. The unknown sample's absorbance is projected on the working curve and the concentration read on the abscissa. A linear curve is preferred because direct calibration is then possible. At higher concentrations the curve may flatten as shown, then further dilution or narrow bracketing of the unknown becomes necessary.

consult the manuals provided by the manufacturers of the equipment for details.

One attempts to work within the optimum analytical range of 20–80% absorption. Nevertheless, a pronounced curvature is often observed in the calibration curve at high absorption. Slit widths and path lengths of burners vary and can even be adjusted in some new burners. These characteristics will, of course, influence the sensitivity. The analytical conditions are given here for general orientation only, an analyst will always adjust and test the conditions before actual data are produced by preparing a *working curve* and bracketing the absorbance of an unknown sample solution as shown in Fig.24.

A linear three-point curve with standards bracketing the "unknown" sample is obtained when *absorbance* is plotted against concentration. In the Perkin Elmer 303 machine, a Digital Concentration Readout (DCR-1) unit, or the use of a chart is necessary to convert the *percent absorption* into *absorbance*. With a linear working curve, a calibration permitting direct readout in p.p.m. saves much time in routine analyses. Readings are taken in the

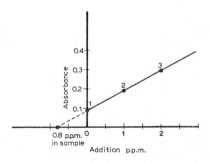

Fig.25. A working curve for the method of additions. Three aliquots of sample are taken, all are diluted to the same volume, first adding 1 and 2 p.p.m. of the metal analyzed to samples 2 and 3. When measured absorbance is plotted in the way shown, the extension of the working curve will intercept the abscissa at a distance corresponding to the concentration of the element in the unknown sample if the same scale is applied, or at 0.8 p.p.m. in this case. The curve must be linear.

sequence standard—sample—standard…. Three to six readings are usually averaged before plotting.

When working with complex mixtures of variable composition, the geochemist often has to use the *method of additions* to minimize the effects of matrix materials. This method consists of the addition of known amounts of the element whose concentration is sought to the sample solution. The plotting of the working curve for the method of addition is done in the manner shown in Fig.25. To avoid changing the matrix substantially, the addition must be restricted to small quantities.

Procedure, method of additions

Take three aliquots of the sample solution, add 1 and 2 p.p.m. of the element to be analyzed to samples no.2 and 3 respectively, and dilute to the same volume. When the absorbance of all three samples is plotted in the way shown, the extension of the working curve will cut the abscissa at a distance corresponding to the concentrations of the unknown sample on the scale chosen. This is true, assuming that the working curve is linear.

Dissolution of samples. Chloride is the preferred anion because it causes minimum interference. Therefore, the metals are dissolved in just sufficient hydrochloric acid, or nitric acid in the case of copper, lead, and zinc. Dissolving aluminum metal in hydrochloric acid requires a droplet of mercury as a catalyst.

The size of the solid sample is usually 100–500 mg, however, to stay within optimum ranges, much smaller samples are often preferred. This avoids excessive dilution. Sampling errors, however, may increase when the size of the sample decreases.

Silicate rocks. Moisten the powder in a platinum crucible with a few drops of water, add 3–5 ml of 72% $HClO_4$ and 10 ml of HF. Evaporate on sand bath under a perchlorate hood to near-dryness. If particles remain undissolved, repeat, otherwise, add 3 ml of $HClO_4$ and evaporate, add 5 ml HCl and evaporate again, repeat the HCl treatment, dissolve in 5 ml of 1/1 HCl, add lanthanum stock solution, and bring up to a desired volume. To prevent hydrolysis of Mg, Ca, Ba, and Sr salts, the addition of equal amounts of NH_4Cl to the sample and standard solutions may be necessary.

Metals and alloys. Dissolve in minimum HCl, or HNO_3 in the case of Cu, Pb, and Zn.

Cement. Disperse 100–1,000 mg of sample powder in 25 ml of water and add 5 ml of concentrated HCl, digest on water bath for 10 min, dilute to 50 ml, digest for 10 min, filter through medium porosity filter, wash the residue well, and bring the filtrate up to desired volume (100–500 ml). Dilute further if necessary.

Sulfides. Decompose in concentrated, nitric acid, aqua regia, or hydrochloric, by digesting the fine powder on water bath; dissolve in dilute hydrochloric, or nitric, if lead is present. The use of hydrofluoric with nitric acid, or the use of perchloric acid with hydrochloric is also recommended when perchloric acid hoods are available. It is always advisable to fume to dryness and dissolve in a known amount of dilute HCl or HNO_3. This assures standard conditions. Remember that with sulfides, sulfate ion will always be present in the solution.

When major elements are analyzed with atomic absorption, it should be kept in mind that the manuals usually give conditions assuring highest sensitivity. This often necessitates dilutions with inherently high dilution errors. The use of less sensitive lines and less efficient flames or excitation conditions, as for example a hollow cathode lamp instead of an Osram lamp for potassium or sodium, may then be recommended. It is often necessary to decrease the path length through the flame. In an extreme case one may turn the burner at a right angle to the light path.

Determination of magnesium

The advantages of the atomic-absorption flame photometry can be demonstrated in the determination of magnesium. As we know from the classical and EDTA methods, given elsewhere in this book, the determination of magnesium in the presence of calcium and vice versa is a rather cumbersome and often not a very accurate procedure unless precautions are taken and great care is exercised. In emission flame analysis, magnesium emits weakly and self-absorption makes calibration at higher concentrations difficult. These are properties, however, which are desirable in atomic absorption.

Magnesium absorbance is affected by aluminum, silicates, phosphates, and sulfates, if present. All these compounds cause suppression of magnesium absorbance due to apparent formation of relatively stable compounds with magnesium in the flame. Alkalies and alkaline earths seem to enhance magnesium absorption in the presence of aluminum by suppressing the formation of magnesium ions, and also by preferential formation of stable compounds with aluminum. Thus, calcium, lanthanum, and strontium salts are often added to counteract the suppression of magnesium by the involatile compounds.

Aluminum and silicon, being the major constituents of silicates, will cause the most serious interferences when magnesium is determined by atomic absorption in these substances. Since silicon can easily be removed by hydrofluoric acid treatment, aluminum, unless removed chemically, remains as the main interfering element.

Recently, a method for magnesium applying a nitrous oxide-acetylene flame has been developed by R.W. Nesbitt at the University of Adelaide, S.A. (Australia). This method seems to have overcome the interference of aluminum on magnesium in atomic absorption analysis of silicate rock samples decomposed by the conventional hydrofluoric-sulfuric acid attack. Nesbitt's method is given in its essential parts below, substituting perchloric for the sulfuric acid, and using less concentrated sample solutions. The interference of alkali metals in this method seems to be negligible in rocks with considerable magnesium, but has to be taken into account in alkali-rich rocks. Calibration and bracketing between equally treated similar samples of known composition is mandatory as always in this type of instrumental analysis. Samples dissolved in different acids require separate working curves. This statement is also true for other elements.

Procedure

Dry 50–100 mg of sample powder at 105°C. Moisten, add 3 ml $HClO_4$, and treat with 5–10 ml of HF. Digest for several hours.

Repeat, if dissolution is not complete, covering the platinum or teflon crucible. Evaporate to near-dryness, add 3 ml of $HClO_4$, evaporate again, and dissolve in 1/1 HCl. Bring to the desired volume, depending on magnesium concentration in sample. To get linear curves, 0.25–2.0 p.p.m. in solution is usually recommended. See that the acid concentration of standards is close to that of the sample solution.

Spray in a nitrous oxide acetylene premixed flame and measure transmittance or absorbance at the 2,852 Å magnesium resonance line. Bracket between selected standards. Repeat six times, rotating between the unknown and the known samples. Convert the transmittance value to absorbance. To check contamination in reagents, run a blank with HCl solution of similar concentration. The range of magnesium determined, expressed in MgO in the sample, is 0.03–13.0%, which covers most igneous rocks, except the ultramafics. When larger concentrations of magnesium are encountered, further dilution is necessary. The sensitivity can also be decreased by turning the stainless steel burner into a right angle position with respect to the light path. Recent work at the University of Adelaide indicates that this method is also accurate for rocks with up to 40% MgO.

Standards for atomic absorption

Standards are prepared before actual use by diluting *stock solutions* of 5,000–500 mg/ml concentration. 50–100 ml of standard solution is made observing that the concentration is within the optimum ranges yielding linear curves. Concentrated, carefully prepared, stock solutions keep better in glass or plastic bottles than very dilute solutions.

When interference of Al, P, Si, and SO_4^{2-} is expected, one adds lanthanum to the sample and standards.

Preparation of lanthanum stock solution. Dissolve 30 g of La_2O_3 in 125 ml of concentrated HCl, and add 375 ml H_2O. Five ml of this solution added to 0, 0.2, 0.5, 0.8, 1.0, 1.5, and 2.0 ml of a 500 p.p.m. Ca standard solution and diluted to 50 ml, gives, respectively,

0, 2, 5, 8, 10, 15, and 20 p.p.m. Ca standard solutions with about 1% La in 5 vol. % HCl. Similar addition is recommended for Sr, Ba, and Mg, if no nitrous oxide oxidizer is used. Some Ca is always present in La_2O_3 as an impurity and the addition of equivalent amounts of this stock solution to the unknown solution is therefore mandatory. Stock solutions and standard solutions can be obtained commercially or prepared from reagents listed in Appendix 2.

CHAPTER 11

Instrumental Nondestructive Analysis

INTRODUCTION

In a strict sense, all analytical methods are instrumental because the balance is the basic tool of a chemist. It is difficult to think of a measurement of chemical or physical properties of an element without the use of some kind of an instrument. Only such properties as color of the substance, visible light emitted, heat, odor, or relative mass, can be directly perceived by the human senses.

Analytical procedures which use an instrument or instruments as a main tool of detection and quantitative estimation are called instrumental methods. Most modern analytical techniques are instrumental, for example, emission, absorption, and fluorescent flame spectrophotometry, spark- or arc-source emission spectrography, potentiometry, electrolysis, chromatography, infrared spectrometry, X-ray emission, and neutron activation. Within this incomplete list of modern instrumental methods, we can distinguish two categories, the so-called *destructive* and the *nondestructive* methods.

Most of the methods above are destructive, by which we mean that in order to conduct the analysis, the sample has to be decomposed by dissolution, fluxing, or melting and burning in an electric current. The sample no longer exists in the original form after the analysis has been completed.

X-ray emission and neutron activation methods, on the other hand, do not require preliminary destruction of the sample in order to excite the radiation that is measured in the detection and estimation process. Thus, these methods are called nondestructive. Naturally, in order to avoid absorption effects and spectral interferences, or when samples are physically heterogeneous, an analyst will sometimes find it necessary to flux and perform preliminary chemical separations when samples of complex composition have to be analyzed by these methods.

ELECTROMAGNETIC RADIATION

In all emission or nuclear activation methods we measure at some stage the electromagnetic radiation which is excited by subjecting the sample to heat or radiation, or accelerated electrons, or bombarding by sub-atomic or other particles. If we present this radiation in terms of wavelength in Ångström units ($Å = cm^{-8}$) on one scale and name the intervals according to established custom, we would read, starting from the short wavelength (or high energy) end: cosmic rays—gamma rays—X-rays—ultraviolet light—infrared light—radar waves—radio or Hertzian waves—and finally the induction and heating waves.

In Fig.26 this *electromagnetic spectrum* is given in relationship to its source the atom, in order to preclude the frequent misunderstanding by beginners who tend to consider somewhat mystically the different segments of this continuous spectrum as separate characteristic types of radiation instead of somewhat overlapping definitions of convenience. For example, an analyst often speaks of hard X-rays meaning gamma-rays, or of soft gamma-rays, meaning X-rays.

What then is this electromagnetic radiation that our sensitive phototubes or photographic plates record, a small segment of which we can see as light or feel as radiating heat? No clear and unambiguous answer can be given to this question because this radiation turns out to be of dual character. We can conceive that radiation must represent *energy alone* as does electricity, gravitation, and magnetism, but when we imagine a total darkness where no radiation exists, we may also think of introducing infinitesimal "particles" of light or radiation. These elemental increments of radiation are called *photons* and treated as *radiation*

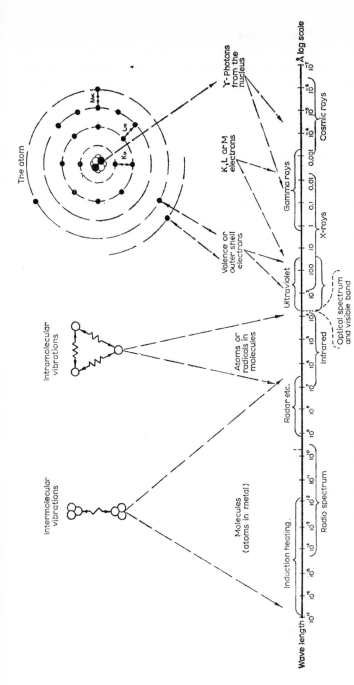

Fig.26. The electromagnetic spectrum and its regions in relation to the molecule and the atom with its electron shell and the nucleus. Simplified and idealized. (Not drawn to scale.)

quanta in distinction of elementary material particles such as atoms, protons, neutrons, or electrons. In order to study the photons we must introduce matter because pure energy can be apprehended only on its interaction with matter.

Planck and Einstein have shown that energy is absorbed or emitted by atoms only in discrete *quanta* or particle-like *photons*. Planck determined this *elementary action constant,*

$$h = 6.62 \cdot 10^{-27} \text{ erg sec}$$

Using the Einstein relation:

$$E = mc^2 \tag{16}$$

while assuming that we deal with pure energy and omitting the mass we can write:

$$E = h\nu = h\,c/\lambda \text{ erg,} \tag{17}$$

where $h = $ *Planck's constant*; $\nu = $ oscillation frequency/sec; $c = $ velocity of light in a vacuum, $2.998 \cdot 10^{10}$ cm/sec; $\lambda = $ diffraction wavelength in cm.

Thus, light or electromagnetic radiation can be treated simultaneously as radiant energy with wave-like characteristics which also has properties similar to that of matter.

Atom as a source of electromagnetic radiation

Energy is a measure of mass and mass is a measure of energy. In order to measure energy we have to study its interaction with matter. We may think, for example, of introducing atoms or other particles into a fictitious vacuum. When the atoms are hit by the photons with sufficient energy, they are excited to emit energy in the form of radiation. Depending on the structure of the atom, the emitted or fluorescent radiation will display a characteristic energy spectrum typical of the atom. The complexity of this spectrum depends on the extranuclear electron configuration and on the energy of the photons interacting with the atomic structure. With the increasing energy of the photons, the electrons are excited to higher energy levels and the electron shell is penetrated deeper so that electrons from inner shells become affected as well.

In terms of the electromagnetic spectrum as compared with an atom's electron shell, we will excite the outer so-called valence electrons by simply heating the sample in a flame. This produces light mainly in the infrared and the visible optical spectrum region. When we burn the sample in an electric arc or spark with their higher thermal energies available, light in the visible and especially in the ultraviolet spectrum region is emitted. In this case, the valence electrons are excited to yet higher energy levels and more electrons may be affected. The light of the optical spectrum is emitted while the excited valence electrons return from the higher energy levels. The complexity of the spectrum increases when we switch from flame to electric arc and spark excitation. In order to produce excitation in this spectral region, we have to generate so much heat that the sample melts and vaporizes partially, thus being destroyed.

Increasing the energy of the photons to levels corresponding with the short ultraviolet (also called ultrasoft X-rays), X-rays, or gamma-rays, results in penetration of these photons through the outer electron shell of the atom and causes interaction with the electrons of the two innermost electron shells of the atom, the so-called K and L shells. The impinging photons knock out these inner electrons from their ground state and *X-ray radiation* is emitted when electrons from outer shells take their place. Using higher energy radiation, we thus penetrate deeper into the atom. Instead of photons, we can also use accelerated electrons to produce X-rays as is done in the X-ray generating tubes. We can thus excite X-rays with X-rays, with gamma-rays, and with electrons.

When we keep increasing the energy of the gamma-rays, we will finally be able to penetrate into the nucleus of some light atoms such as lithium, beryllium, or boron, and cause equilibrium disturbances within this core of the atom, which, as we know, consists of protons and neutrons. When a gamma photon or some sub-atomic particle causes disturbance within the nucleus of an atom, we call the process *nuclear activation*. In order to penetrate the electron shells of heavier atoms, accelerated *neutrons* are mostly used, but also electrons (e) or beta particles (β), protons ($^1H^+$), deuterons ($^2H^+$), tritons ($^3H^+$), and helions ($^4He^{2+}$, $^3He^{2+}$, also called alpha particles) can be accelerated. Neutrons, being neutral particles, pass through the shell of negative electrons and combine with the positive nucleus easier than the charged particles listed above. The process is called *neutron activation*. When neutrons react with the nucleus, gamma (γ) radiation, beta (β) radiation, or alpha (α) radiation, or a combination of these radiations may result from the product isotope. It is usually the gamma radiation that is measured in order to identify the atom or atoms involved and to estimate their concentration in the sample. The process is called *scintillation or gamma ray spectrometry*.

From the above, we conclude that with the increasing energy (shorter wavelength) of a photon, we can penetrate deeper into the atomic structure and that the excited radiation will accordingly be of shorter wavelength. We can summarize by stating that photons, emitted by disturbed valence and other electrons from outer shells, constitute the *optical spectrum*. Photons emitted by the displacement of the inner K, L, or M shell electrons are conventionally called *X-rays*. When we use neutrons or other nucleons as projectiles, the process is called activation and the photons coming from the disturbed radioactive nucleus are called *gamma-rays*.

Most analytical instrumental methods described in this book consist of measuring the specific energies of the emitted photons, irrespective of whether these have been derived from photon interaction with extranuclear electrons, or from nucleon interaction with the nuclei of the atoms. This characteristic radiation identifies the atom, and when its intensity is measured, it permits us to estimate the relative amount of the element present in the sample.

Separation, detection, and measurement of the electromagnetic radiation

Various devices are used to separate, detect, and measure radiation, depending on the energy of the photons. Conventional notations of the different spectral regions originate from the practical limiting factors caused by the different dispersion, detection, and measuring devices which are efficient only over specific wavelengths, or energy intervals of the whole electromagnetic spectrum.

To *separate* the radiation in the *optical spectrum range*, monochromators, prisms, or gratings are used. To *detect* the radiation dispersed by these means, light-sensitive electronic tubes and photographic paper or plates are employed. Because, however, optical spectrography is mainly applicable in trace element analysis, and because more detailed descriptions of principles of flame emission and atomic absorption were given in other chapters, it will not be further treated here.

In order to separate X-rays according to their specific energies, diffracting crystals are used (p.238).

The *detection of X-rays* can best be accomplished by an array of proportional counters, *Geiger* tubes, and scintillating crystals. The choice depends on the energy of the X-rays measured. For "ultrasoft" X-rays ($\lambda = 100\text{--}10\text{Å}$) one uses flow proportional counters, for "soft" X-rays ($\lambda = 10\text{--}3\text{Å}$) proportional counters or Geiger tubes, and for "hard" X-rays ($\lambda = 3\text{--}0.1\text{Å}$) mostly scintillation counters. Naturally, the use of these detectors overlaps.

In order to facilitate the understanding and orient the beginning chemist, Table VII lists the dispersive media, and the detector types used to measure radiation in the different regions of the electromagnetic spectrum. It also gives the frequency, ν, of oscillation and the corresponding wavelengths in meters or Ångström units. When X-rays and especially the gamma-rays are measured, one also frequently uses the unit of energy called *electron volt*, abbreviated eV. One electron volt (1 eV) is equal to an energy acquired by an electron falling through a potential difference of one volt. The photon energies within the X-ray spectrum from lithium to californium vary from about 50 eV to about 130,000 eV. The photon energies of the gamma-ray spectrum spread between about 10,000 and about 16,000,000 eV.

In gamma-ray spectrometry, one usually speaks in units of millions of electron volts, abbreviated meV, which are more convenient to use. The reader will notice that in Table VII and in the spectra described above, the wavelengths, frequencies, and energies

TABLE VII

DEVICES USED TO DISPERSE AND DETECT DIFFERENT REGIONS OF ELECTROMAGNETIC RADIATION

Type of radiation	Wavelength	Frequency cycles (sec^{-1})	Dispersive medium	Detector	Cause of excitation of atom
Radio spectrum, television, and radar	$<30,000$–0.001 m	10^4–10^{12}	combination of regulated capacitors and tuning coils	antenna, oscillating crystal loudspeaker, fluorescent (television) screen	vibration of oscillator
Infrared spectrum	$\sim10^{-3}$–$\sim10^{-6}$ m	$\sim10^{12}$–$\sim10^{14}$	special prisms, KBr, CsBr, gratings,	photosensitive tubes, special films	thermal excitation
Optical spectrum, including visible light, overlapping into infrared and ultraviolet	$10,000$–$\sim2,000$ Å	$\sim10^{14}$–$\sim10^{15}$	glass, quartz, and fluorite prisms, gratings	photografic film or photo-tube	flame, electric arc and spark, light
Far ultraviolet	$2,000$–100 Å	$\sim10^{15}$–$\sim10^{16}$	fluorite prisms, gratings	special photo-tubes, flow proportional counters	electric arc and spark in vacuum
X-rays	100–0.1 Å	$\sim10^{16}$–$\sim10^{20}$	diffraction crystals and pulse height analyzers	flow proportional (and Geiger) counters, scintillating crystals	X-rays, accelerated electrons, gamma-rays
Gamma-rays	1.0–10^{-3} Å	$\sim10^{19}$–$\sim10^{22}$	electronic differential pulse height selectors and multichannel analyzers	scintillation crystals, solid state detectors	natural radioactivity, accelerated neutrons and other particles
Secondary cosmic rays	10^{-3} Å	$>10^{22}$	electronic discriminators and analyzers	scintillation crystals and liquids	cosmic rays

partially overlap. This is done on purpose to emphasize the fact that in analytical work one identifies the radiation according to the source rather than according to strict energy intervals.

Speaking of nondestructive methods, we can restrict this chapter on measurement of electromagnetic radiation to the X-rays and the gamma-rays.

The *Geiger-Müller counter* tube consists of a cylindrical cathode with a central anode wire, both placed in a tube usually filled with a mixture of methane and argon gas at pressures below atmospheric (Fig.27). When a potential is applied, usually between 800–1,500 V, and an ionizing gamma photon, a beta, or an alpha particle, happens to enter the field between the electrodes through a thin window, and collides with an atom of the filling gas ejecting an electron, an electrical discharge or avalanche occurs which can be measured electronically as a pulse. In G-M counters, the size of the pulse is nearly independent of the energy of the photon. Thus, while efficient counting of gammas, betas, and alphas, can be performed with a G-M counter, the distinction of different types of radiation can not be achieved.

A *proportional counter* is similar in construction with the G-M counter, but usually uses a lower voltage from 500 to 800 V. In this region, the number of ion-pairs produced by the secondary electron depends on the energy of the photon entering the gas chamber. Photons of higher energy produce electrons that travel faster, causing more ion-pairs to be formed. We speak of *ion-pairs* because each time an electron, or a charged particle, collides with an atom and knocks out an electron, thus forming a positive ion, the

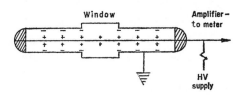

Fig.27. A cross-section of a Geiger-Müller tube showing the principle of operation.

liberated electron may attach itself to another atom forming also a negative ion. The positive ion and the free electron are also considered as an ion-pair. The more ion-pairs are produced, the stronger the pulse will be. When these pulses are *sorted electronically* according to their energy, we may distinguish between different photons and identify the specific gamma or X-rays according to their energies.

In the *proportional region* when the potential gradient near the anode is high enough, the electrons produced by the photons acquire enough speed on their way toward the positive electrode to cause the formation of additional ion-pairs and electrons, which in turn speed toward the anode. This chain reaction reaches avalanche-like proportions, causing amplification which is a characteristic of the type of proportional counter. This so-called *gas-amplification factor* may be as large as 10^6 in some proportional counters and 10^8 in G-M counters, and it is an important characteristic of these counters because stronger pulses are easier to record and amplify electronically. The pulses so recorded remain proportional to the energy of the intial photon or charged particle as long as the voltage of the proportional counter is kept constant.

Flow proportional counters frequently used in detection of ultra-soft and soft X-rays operate on principles similar to proportional counters with the difference that a steady gas flow through the counting chamber is maintained. The gas used is usually a mixture of 10% methane and 90% argon. After flushing, the flow rate is usually maintained at about 10 cm^3/sec in these counters.

To properly operate the different types of counters described, one selects the specified *high-voltage* region, and determines the *plateau* in Geiger-type counters. The plateau is the region where the change of the voltage does not affect the pulse rate. It usually extends over a range of 200–300 V. An *operating voltage* somewhat below the middle point of this range is usually selected because minor voltage fluctuations, common to all electronic equipment, do not then appreciably affect the pulse counting rate (Fig.28).

The *pulse size* or energy naturally does depend on the applied voltage in these counters, as shown in Fig.29, which also demon-

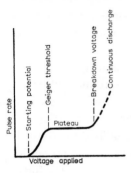

Fig.28. A diagram indicating the high-voltage "plateau" region of a Geiger-Müller counter. In this region, small voltage changes do not affect the pulse rate.

strates the relative position and application of the different gas ionization counting ranges in terms of the high voltage applied. In the *ionization chamber* voltage range, we notice a plateau where the pulse size is little affected by the potential applied, however, this range is not useful in gamma-ray counting due to the weak pulses which are insufficiently amplified within the chamber.

In X-ray analysis the photons measured by the proportional counters are normally spread over a relatively narrow energy range. This is achieved by diffracting crystals which can be rotated in a

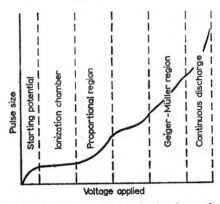

Fig.29. A diagram showing the dependence of pulse size on the applied voltage demonstrating the different counting regions employed.

goniometer so that the incident X-ray beam from the sample meets the crystal at a desired angle, and the detector is placed so that it intercepts only a narrow part of the diffracted wavelengths (p.239). Thus, if the dispersion is adequate, one could simply use a Geiger-Müller counter to measure the bulk of the radiation reaching the counter. For most X-ray emission spectroscopic applications, this degree of discrimination is not sufficient, and, therefore, the proportional counters are utilized to give additional refinement in the separation of X-ray quanta. The proportional nature of the signals recorded by these counters permits further sorting by electronic means according to the energy of the pulses. This process is called *electronic discrimination or pulse height analysis*. Electronically, it is possible to accept or reject pulses with greater or smaller amplitudes than a set value. It is also possible to select the *threshold* and the *upper limit* to apply simultaneously so that only pulses of a given narrow energy distribution are accepted. We speak of a *window* in the latter case, which means that we have selected a narrow range of the total photon energies entering and activating the counter to be measured and recorded. In this case, we plot the amplitude corresponding to the energy of the radiation along the abscissa and the frequency of pulses or the total counts recorded on the ordinate. This can be done electronically and recorded as shown in Fig.30. We can either *integrate* the counts under the peaks by moving the threshold or *differentiate* by moving a narrow electronic window over the energy region of the radiation reaching the counter.

One often detects an additional peak on the lower energy (longer wavelength) side of the radiation peak measured by proportional gas-filled counters. The radiation causing this peak is produced within the counting chamber by collision of the entering photons with the atoms of the counter gas. If the energy of the photons is sufficient, a K-shell electron is knocked out, producing a corresponding X-ray quantum, most of which usually escapes the counter. The initial energy of the photon is reduced, however, by this X-ray quantum and is recorded at a correspondingly lower amplitude as an additional peak called the *escape peak* (Fig.30). This

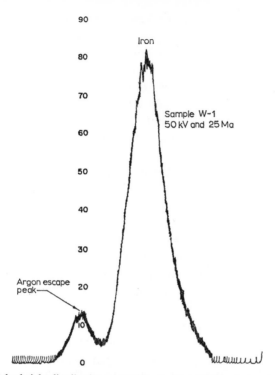

Fig.30. Pulse height distribution peak of iron with an argon escape peak on the low-energy side. Flow proportional counter 90% Ar + 10% CH₄, high-voltage 1,500 V, window-voltage 1 V, crystal EDDT. (Courtesy of B. P. Fabbi.)

name is confusing because it is not the energy of the escaping radiation, but the reduced energy of the photon which initially had entered the chamber, that is measured. The distance of both peaks will depend on the difference between the original photon energy and the X-ray quantum characteristic of the gas used. This means that the escape peak will occur at an energy level corresponding to that of the gas used, and both the escape peak and the photon peak corresponding to the element measured will be separated by the same voltage gap as are the main peaks of these elements. If the original photon quantum is below the excitation potential of the element forming the counter gas, no escape peaks will result.

Therefore, heavier noble gases such as xenon or krypton are often used as fillers for this purpose. To understand this, one must know that the minimum energy necessary to knock out a K electron with another electron or a photon increases with the atomic number. This energy is a characteristic of each element and is called *K absorption energy*, conventionally referred to as the *absorption edge*. Analogically, we may speak of the L absorption energy. Expressed in terms of wavelength, these are often denoted as $\lambda \, K_{abs}$ or $\lambda \, L_{abs}$.

Scintillation counters

The scintillation counter is perhaps the most widely used detector of hard X-rays and gamma-rays. When the energy of the photons increases, approaching the gamma-ray region, the gas proportional counters described above become less efficient, and we must use specific crystals in which photons produce microscopic flashes of light upon collision with the atoms. The oldest device of this type was used by Crookes in 1903, who detected in a darkened room under microscope, flashes on a zinc sulfide screen, produced by alpha particles emitted from a radioactive substance. Pioneering physicists in radioactivity performed their early work counting visually these flashes of visible light called *scintillations*. The first "manual" ZnS counters were called *spinthariscopes*. The restricting factor in the use of these counters at that time was the absence of sensitive detection devices to measure and amplify these weak light pulses, and the above-described gas ionization chamber and Geiger-Müller counters soon displaced the scintillation counting.

It was not until the development of the sensitive *photomultiplier tube* (Fig.31) after the Second World War that the modern scintillation counting became feasible. This discovery, with the resulting need for better scintillating materials or *phosphors*, has led to the discovery of a group of highly effective organic and inorganic crystalline and liquid phosphors by about 1950.

In the *inorganic group* of these phosphors, thallium-activated (doped) sodium iodide, NaI(Tl), is still the most frequently used

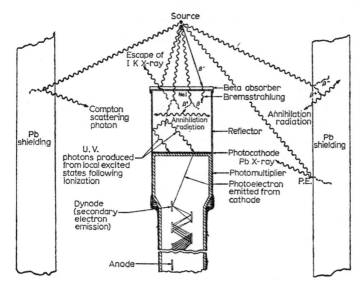

Fig.31. A schematic of a sodium iodide scintillation detector, a photomultiplier tube, and some of the events taking place in the detector, its immediate vicinity and shielding materials. (After HEATH, 1964.)

crystal, but other activated (doped) alkali iodides, CsI(Tl), CsI(Na), KI(Tl), LiI(Eu) and a europium-doped calcium fluoride, $CaF_2(Eu)$, have been developed. Also, glass scintillators are known. In the *organic group*, crystals and organic solid and liquid solutions are employed. Napthalene was one of the earliest used in the organic crystal group, now anthracene and stilbene have replaced it. Organic solutions for the same purpose consist of plastics (lucite, solid) and liquids which have the advantage that the sample counted may be totally immersed into the counting medium. The most important characteristics of scintillators are:(1) the *wavelength of maximum emission* which is usually in the visible spectrum region or in the near-ultraviolet; (2) the *decay constant* of the pulse which is in the 1-microsecond (μ sec = sec^{-6}) range for the inorganic crystals and in the nanosecond (n sec = sec^{-9}) range for the organic crystals and solutions; (3) the *density* which varies in the region of 3 and 4 g/cm^3 for inorganic and is near 1 g/cm^3 in the organic scintillators; and (4) the *relative pulse height*, which varies by a factor of 8 (1–8×) with the inorganic crystals. Especially the NaI(Tl) crystal gives relatively high pulses which are consequently easier to amplify.

With most of the scintillators listed, one can count gammas,γ, betas, β, and alphas, α. The organic scintillators are also useful for high-energy or fast-neutron counting, and especially the NaI(Tl) crystal is suitable for counting X-ray quanta.

When pulse height analysis of counted X-ray quanta is performed in X-ray emission spectrography, electronic *single channel pulse height analyzers* are used (p.216). This is possible because the recorded energies vary only within a relatively narrow energy band due to diffracting crystals employed in the first step of dispersion. If one omits the analyzing crystal, all X-ray energy quanta originating from the sample and reaching the counter would be recorded as pulses. This would mean recording over the total X-ray band width or over some 100,000 eV, simultaneously. In order to do this electronically in a single channel analyzer with required resolution, a prohibitively wide amplification range would be required, even if scintillation crystals or gas counters with such resolution could be found. Also, admitting all pulses to be counted may crowd the amplifier causing *dead time* problems, because no matter how fast a pulse decays it takes an infinitesimal but measurable period of time to be counted, during which time the counter is incapacitated to accept or record other signals. With increased frequency of pulses, an increasing amount of signals will thus remain unaccounted for.

In order to accomodate a wide energy band of pulses, scintillation crystals with phototubes and electronic devices consisting of a great number of joined single channel analyzers are constructed. Each of these single channels is set to accept only pulses of a narrow energy range. These devices are called *multi-channel analyzers*. With these instruments, the entire gamma or X-ray energy spectrum can be recorded at one time. They may be used to differentiate the counter pulses directly without preliminary dispersion media between the sample and the counter. An X-ray emission analysis performed employing multichannel analyzers is called *nondispersive* in reference to the absence of the diffracting analyzer crystal. The exploitation of multichannel analyzers in nondispersive X-ray emission work is presently in its early development stages, whereas their primary uses have been, and are in, the gamma-ray spectrometry. Magnetic memories and computer-like adding and subtracting, integrating and differentiating circuits are integral parts of modern multichannel analyzers. Units with one hundred channel memories, up to units capable of handling and sorting information from thousands of channels have become recently available.

When scintillation crystals are used in connection with multichannel analyzers, it is important to know the *resolution* of the crystal which varies with gamma energy, and hence must be defined for some particular gamma. By definition resolution is measured on the ^{137}Cs line. Compared to the resolution in the low-energy X-ray field achieved by diffracting crystals, this resolution is less by orders of magnitude. For example, what would be called a relatively good resolution of the 0.663 meV peak of radioactive cesium-137 (^{137}Cs) is some fifty kilo-electron volts, 50 keV, which corresponds to roughly one half of the total band width of the X-ray spectrum between lithium (~ 55eV), and uranium (115 keV).

Whereas resolution in optical spectrometry is expressed in the same units as used describing the spectral line, or in Ångstroms for example, the resolution of a scintillation crystal is calculated in relative *full line width at half maximum height* of the peak (F.W.H.M.). This characteristic of the crystal is derived by dividing the width of the energy distribution peak at half maximum height, ΔE, by its average energy, $\overline{E}$, and it is expressed in percent as shown in Fig.32. One speaks of *percent resolution* of a certain crystal at a given average energy of the gamma peak. In the case demonstrated in Fig.32, one calculates:

$$\frac{53 \text{ keV} \cdot 100}{663 \text{ keV}} = 8.0\%$$

The relative quality of crystals is conventionally determined by

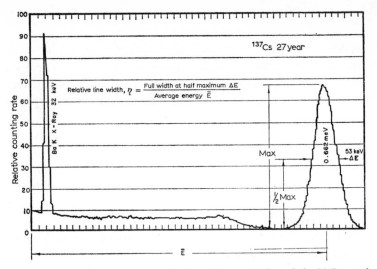

Fig.32. ^{137}Cs spectrum and the calculation of the resolution of the NaI crystal used as detector. (After W. C. Kaiser, *Anal. Chem.*, 1966 p. 29A.)

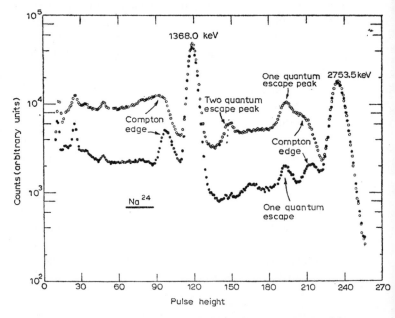

Fig.33. The effect of anticoincidence counting on Compton scatter and the escape peaks. ● = with anti-coincidence; ○ = without anti-coincidence. (After W. C. Kaiser, *Anal. Chem.*, 1966 p. 33A.)

comparison of their resolution at the [137]Cs 0.663 meV gamma peak. As demonstrated in Fig.33, gamma-ray spectra usually show a *continuous spectrum* which is due to *Compton scattering* and *escape* of electrons from the crystal. Compton scattering results from electrons caused by photons which lose only one part of their energy through inelastic scattering after collisions with the atoms of the crystal. The loss of energy by the photon, and consequently the energy of the electron, will depend on the angle through which the photon recoils. The Compton scattering is recorded as a continuous distribution of pulses on the low-energy side of the photometric peak. In addition, there often appears a somewhat indefinite peak called *Compton peak* or edge (Fig.33) due to 180° recoil of photons.

The escape of electrons from the crystal before producing the maximum of ion-pairs proportional to the energy of the incident photons also contributes to this continuum. The effect of the Compton continuum and the escape peaks can be minimized by *anti-coincidence* counting (Fig.33). An anti-coincidence spectrometer usually has a small crystal within a large crystal and accepts only the pulses of the central crystal when they are not accompanied by pulses from the surrounding crystal which counts the escaped photons. When coincidence occurs, the pulses are electronically discarded.

The main energy distribution peak, called the *photopeak* (at 662 keV in Fig. 32) is caused by the recoil of electrons after collision with the incident gamma photons. In this so-called photoelectric interaction, all the energy of the photon is absorbed by the electron. The kinetic energy of the electron will then be equal to the energy of the incident photon minus the energy required to ionize the atom. The photon is thus annihilated, and the recoiling photoelectron, while colliding with the atoms of the crystal, produces a series of ion-pairs proportional in number to the initial energy of the incident photon, unless it happens to escape the crystal before expending all its energy. This will naturally happen with increasing probability when the size of the crystal becomes smaller and the energy of the incident photon increases. Thus smaller crystals as well as increasing gamma-ray energy will cause an increasing lower energy background, often referred to summarily as the "Compton background" even though it is produced by the escape of electrons as well as the Compton scattering.

While following the discussion above, the reader must remember that in the scintillation counter we are measuring not the direct multiplication effect of the ion-pair production as in the gas proportional counters, but the light flashes which are proportional to the energy of the incident photons. These scintillations are produced by the re-emission of a small part of the energy which is absorbed by the atoms during the collisions with the photons and electrons, as light (photons).

When the energy of incident photons exceeds 1.02 meV, a process called *pair production* becomes possible. In this case, a photon converts into a positive electron (positron) and a negative electron (negatron). This process becomes more significant when the initial photon energies are above 3 meV and it is usually recorded as two additional peaks on the low-energy side of the photopeak. These are called *escape peaks* and they are located at energies less by 0.511 meV and 1.02 meV than the photopeak.

The appearance of two additional peaks at constant energy intervals can be understood when we consider the formation and interaction of the positrons and the negatrons. It has been observed that in order to transmute a photon into two "electrons" of opposite charge, the incident energy has to be above 1.02 meV. It has also been recorded that when a positron collides with a negatron both disappear or are *annihilated*, while two gamma-ray photons, each of 0.511 meV energy and moving in opposite directions (at 180° angle) are formed. Returning to our crystal, we can now imagine a high-energy photon producing a positron-negatron pair, in which process 1.02 meV of the energy is spent, while the rest of the incident photon energy is transmitted as kinetic energy to each member equally. When annihilation occurs, two 0.511 meV photons are produced per pair. If one of these photons escapes the counter while all the other events have produced a photoeffect, a *single escape peak* will appear at 0.511 meV less than the photopeak. If both of the annihilation photons escape detection, a second so-called *double escape peak* will be seen. These peaks are seen super-imposed over the Compton continuum of a sulfur-37 gamma-ray spectrum in Fig.34.

The *complexity* of the gamma-ray spectrum is often further in-creased by the appearance of the "annihilation peak" at 0.511 meV, the "backscatter" peak, and the "sum peak". The annihilation peak which always appears at 0.511 meV when recorded, is due to pair production in the shield of the detector. When these photons enter the crystal, they are counted with the other radiation. The back-

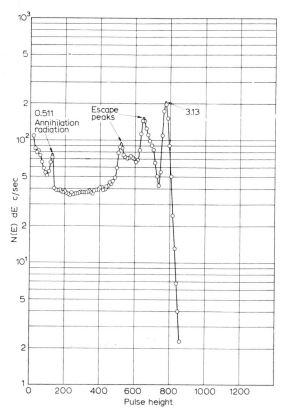

Fig.34. Gamma-ray spectrum of 5 min ³⁷S isotope showing the two escape peaks and the annihilation radiation peak. 3 × 3 inch −2 NaI; absorber 1.34 g/cm²; source distance 10 cm (c); energy scale ≃ 4 keV/P.H.U. (After HEATH, 1964.)

scatter can be due to Compton scatter or characteristic X-rays, for example, lead X-rays produced by gamma photons on interaction with shielding material of the crystal (Fig.31). The sum peaks are produced through coincidence of two so-called "cascade" gamma-rays being counted simultaneously with a total loss of energy in each case. Cascade gammas are sometimes emitted in connection with beta decay. For example, the 5.3 year cobalt-60 isotope emits two concurrent gammas at 1.17 and 1.33 meV energy. The resulting "sum" peak, when detected, thus appears at 2.50 meV.

Also, the effect of radioactive background has to be taken into account. Natural radioactive isotopes ^{40}K, uranium, thorium, and their radioactive decay products (daughters), especially the radon gas, and also ^{138}La, ^{147}Sm, and ^{87}Rb, can all interfere. In addition to the relatively complex nature of a single gamma spectrum, the nuclear spectroscopist must often analyze mixtures of radioactive isotopes.

Spectrum stripping helps in the interpretation of complex gamma-ray spectra and it is a common procedure in gamma-ray spectro-metry when multichannel analyzers are used in combination with integrating equipment. This technique consists of computer-controlled subtraction of certain spectrum components representing a known gamma-ray spectrum of known relative intensity from a composite spectrum. This is done in order to better identify the unknown component and measure its relative gamma-ray intensity. Fig.35 demonstrates this method of analysis of activated samples in a simplified schematic.

For example, when radiation of two radioactive isotopes in a sample is measured and recorded in the memory of the multi-channel analyzer, we get a composite gamma-ray spectrum shown. Three major peaks are seen, of which one corresponds to the known energies of the isotope M, and two peaks belong to the unknown isotope X. We want to measure the relative concentration of this nuclide by integrating the total counts under the peaks. If we do that directly, the interference of the isotope M, with resulting background, will cause inaccuracies in our determination. If we could subtract the known component of the spectrum, the relative total

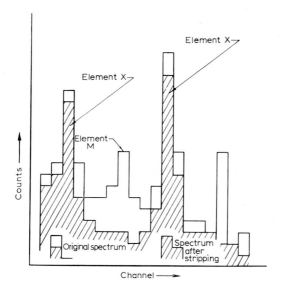

Fig.35. Spectrum stripping. Two superimposed gamma-ray spectra of elements M and X form a composite spectrum. If we subtract the spectrum of the known isotope M electronically till its peak disappears, we will be able to better identify and estimate the isotope X.

counts under the peaks of the unknown would correspond better to the true concentration of that isotope and would also be more recognizable in the qualitative sense. In multichannel analyzers with memories divided into two or four parts, we could now accumulate the known spectrum from a pure sample of known composition for the same period of time into one memory half and then subtract it from the complex spectrum memorized in the other half. This results in simplification and correction of relative intensities under the peaks, as shown in Fig.35. In modern multichannel analyzers-integrators, one can subtract or add selected percentages of known spectra consecutively, which permits the analyst to detect when peaks of known components disappear and to evaluate the relative effect of this contamination.

Semiconductor detectors

The complexity of a gamma-ray spectrum and the relatively poor resolution of crystals limit the applicability of phosphors in multi-channel analysis. We have seen that the resolution in terms of energy in electron volts in the meV region is such that about one half of the total "energy spread" of the elemental X-ray spectrum corresponds to the relative half width of a photopeak, which means that if we work with a 99% confidence, just about the whole elemental X-ray spectrum will fit within these limits. In actual analytical work, this poor resolution of scintillators often necessitates chemical pre-treatment and separation of elements in activated samples, thus seriously impairing the analyst's ability to detect rapidly decaying isotopes and to perform a truly nondestructive activation analysis. Confronted with these limitations, the nuclear spectroscopist has sought detectors with better resolution.

In recent years, so-called lithium drifted germanium and silicon semiconductor detectors have become commercially available. These detectors show resolutions up to two orders of magnitude better than the sodium iodide crystal. They have broadened considerably the application of nondestructive nuclear activation analysis. Due to the scope of this book, only a very incomplete explanation of these devices can be given.

Because we consider here gamma detection primarily, this short description will be mainly confined to the lithium drifted germanium detector. Germanium, $Z = 32$, has sufficient stopping power for the high energy photons, whereas silicon, $Z = 14$, is more suitable for nuclear particle detection.

Solid state counters are ionization chambers in which radiation is absorbed releasing certain charges which can be electronically measured. A photon quantum, for example, produces secondary electrons when it is stopped or scattered in the intrinsic region of the counter. These secondary electrons cause further ionization in a cascade like process until they expend their energy sufficiently to be trapped by "holes" (locations lacking an electron) which exist in the crystal structure of the detector. The number of ion pairs (electron hole pairs) produced along the path of the secondary electron is proportional to the energy deposited by the primary photon or radiation. Hence, a linear relationship results between the signal amplitude and the energy of the incoming radiation above a certain low threshold level (3.6 eV in Si; 2.9 eV in Ge). In the low average energy needed to form an ion-pair lies the main advantage of the semiconductor detectors when compared to the gas ionization chambers (~ 30 eV) and scintillator-photomultiplier combinations (~ 300 eV per photo electron). The greater density of solid detectors permits complete absorption of energetic betas and protons in millimeter thicknesses compared to meterlong ranges in air or gas.

A *lithium-drifted germanium semiconductor detector* is usually a waferlike package consisting of a thin layer ($< 1\,\mu$) of a p-type material, an intrinsic thicker lithium-compensated germanium layer (1–5 mm) and thin n-type layer. These wafers can be stacked for better stopping power of high-energy gammas.

A single detector of this type is actually a large-area junction-diode with two electrical contacts made through vacuum-evaporated thin gold film on the p-layer and an ohmic contact with the thicker semiconductor layer. When a certain potential is applied to this system, charge carrier particles will move in the direction of the potential gradient. When ionizing photons or some charged particles enter this system and produce ion-pairs, minute voltage fluctuations can be recorded as measurable pulses. These pulses are proportional to the number of ion-pairs produced by the secondary electron. When voltage is applied to this diode, a relatively small amount of energy is required to form an ion-pair as seen above. This results in a narrow energy distribution of the pulses because the signal is correspondingly stronger per unit of photon energy which results in better statistics and consequently in better resolution, as compared with the much broader energy distribution of the light flashes in a scintillation counter.

The n- or p-type layers are made by "doping" or diffusing certain metals and non-metals into a thin outer layer of the semiconducting substance. The main purpose of diffusing lithium into germanium is to obtain higher resistivity. Lithium being an electron donor fills the interstitial sites tending to pair with impurities of an opposite or electron acceptor type. Collectively, these impurities then make no contributions to the free carrier concentrations. Doping becomes necessary in order to improve the characteristics of the ultra-pure substances used for detection purposes because of practical limits of purification attainable by zone melting techniques.

Unfortunately, in order to be able to collect or measure the total effect of the ion-pairs formed in a semiconductor detector, these devices have to possess a certain thickness. Due to technical difficulties in building deeper depletion layers for these devices, the thickness of the layer or layers employed is severely limited, which results in relatively lower counting efficiency when compared with phosphors, nevertheless, large volume coaxial solid detectors have been built.

Another difficulty lies in the fact that the ion-pairs produced within the germanium semiconductor layer are trapped quickly at room temperature if the interstitial lithium is permitted to drift away, thus escaping from being recorded. This necessitates working at reduced temperatures, usually with liquid nitrogen at $78°K$, and in a vacuum for better insulation. The life of these detectors at room temperature is about one hour. These devices are also light-sensitive and must be placed in the dark. A typical semiconductor detector system for high resolution nuclear spectrometry is shown in Fig.36.

Ultra high purity germanium or silicon are required to keep the number of free charge carriers at a minimum so that only the photon-induced ionization is counted, thus lowering the background noise. The lattice of these substances, however, will have a certain electrical conductivity in the field applied, which causes this current or noise.

Even when all these precautions are taken, the useful life of these detectors is presently about six months to a year. The range of electron energy covered by these detectors stretches from about 20 keV to above 10 meV, depending on the thickness of the detector. Fig.37 shows gamma-ray spectra of some isotopes recorded with a NaI(Tl) crystal and a liquid nitrogen cooled semiconductor detector.

Fig.36. A typical semiconductor detector system. (Courtesy of the Technical Measurement Corporation.)

Resolutions achieved with systems such as this are in the 1–2 keV range.

As in the scintillation counter, three primary processes are distinguished when gamma-rays interact with the Ge(Li)-detector. These are: (*a*) photoelectric effect, (*b*) Compton scattering, and (*c*) pair production (p.223). If all the photon energy is expended within the active volume of the detector, the signal obtained will contribute to the full energy peak (photopeak). Thus the *photoelectric effect*, produced when the entire energy of the photon is absorbed, is not the sole contributing factor to the photopeak recorded in the pulse-height spectrum. Multiple processes, listed

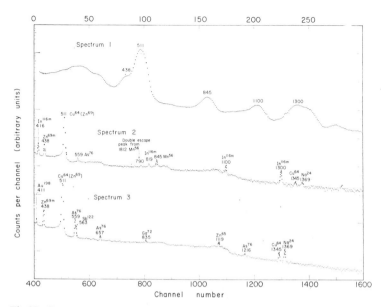

Fig.37. Gamma-ray spectra of activated sulfide ore. *Spectrum 1*: 60-min data accumulation with a 3 × 3-inch NaI(Tl) crystal. 5.2 h, after a 2-h irradiation of 62.9 mg of sulfide ore. *Spectrum 2*: 60-min data accumulation with lithium-drifted germanium detector, Ge(Li), 4 h after a 2-h irradiation of 62.9 mg of the sulfide ore. *Spectrum 3*: 120-min data accumulation with Ge(Li) detector 24 h after a 49-h irradiation of 50 mg of the sulfide ore. (After Lamb. Prussin, Harris and Hollander, *Anal. Chem.*, 1966, p. 814.)

above, contribute significantly to this peak, and may in fact be dominating in some cases.

The collection of the free charge carriers in a solid detector is performed by impressing a voltage gradient of 50–100 V/mm over the intrinsic region of the crystal. The resultant current can be electronically integrated, amplified, and signals sorted according to their amplitudes in multichannel analyzers. An energy spectrum thus produced, when calibrated, permits the study of the incident gamma radiation. Such spectra consist of one or more narrow peaks superimposed over the continuum (Fig.37). A gamma-ray spectroscopist is usually interested in the determination of the height of these peaks relative to the continuum. The position of these peaks on the x-axis of the recorder, or their *pulse height* is related to the energy of the photons and permits the identification of the emitting source or atom.

Selecting a Ge(Li) detector for specific experimental purposes, one considers primarily (*1*) the *energy resolution* required, (*2*) the *counting rate* expected, and (*3*) the *time resolution*.

In an ideal germanium detector:

$$\text{F.W.H.M. (keV)} = 1.44 \ \sqrt{\overline{E}} \ \text{(meV)} \qquad (18)$$

where F.W.H.M. denotes the full width at half maximum height (p.221) in kilo electron volts, and E is the photon energy in million electron volts.

When the thickness (volume) of a detector with a fixed area is increased, the *counting rate* is improved but a higher bias voltage is required to collect the charged carriers efficiently. Incomplete collection results in asymmetric peaks which impairs the *energy-* and the *time resolution*. When voltage is increased, however, leakage currents also increase raising the background, which again affects the resolution adversely.

Increasing the collecting electrode area with fixed thickness magnifies the capacitance of the detector. Coupled to a low-noise pre-amplifier, the noise of such a device increases with its size. This reduces the peak height relative to background which is especially undesirable when lower photon energies are measured. Consequently, each counting problem requires an optimized detector which is chosen to give maximum energy resolution at a given full-energy peak combined with maximum attainable counting rate. The *time resolution* is generally limited by the electronic circuitry only.

These requirements set practical limits on the capacitance (size) of the detector and its thickness. Low-noise pre-amplifiers must be used. In general, low-detector capacitance (small size) will give better energy resolution. *Counting rate* is increased, however, by added thickness or volume. When the energy of photons increases, pair production and with it escape peaks and the Compton background also increase which may mean lower peak to background ratios for the photopeak measured. The geometry of the counters (five-sided coaxial and cylindrical detectors are now available commercially) and the position of the source also have an important bearing on the efficiency of a detector–pre-amplifier–amplifier–multichannel analyzer assembly.

CHAPTER 12

X-Ray Emission Analysis

INTRODUCTION

Following the discovery of X-rays by Röntgen in 1895, the basic principles of X-ray spectrographic analysis were developed in the first two decades of this century. After the relation of the characteristic X-ray radiation to the atomic number was established by Moseley, Von Hevesy and other workers applied this radiation to the analysis of minerals and rocks. Concentrations of different elements down to one part in one hundred thousand parts were measured, and the advantages of this method were uniquely demonstrated by the discovery of the element hafnium.

Despite a very promising and successful start, the widespread application of X-rays to general chemical analysis has developed only in the period of approximately the last fifteen years. This delay was caused by the high experimental demands of the technique as first applied with demountable tubes and samples used as targets. The main factors which helped to make X-ray emission analysis a universal method were the development of: (*1*) high intensity sealed X-ray tubes and stable power sources; (*2*) highly sensitive Geiger, proportional and scintillation counters; (*3*) large crystals with high diffracting power as dispersion media; (*4*) reliable, precise, and fast electronic counting apparatus and monitoring systems; and finally, (*5*) the development of helium and vacuum path spectrographs. This latter development has extended the range of X-ray analysis to include light elements to atomic number 12. Important advances in the X-ray techniques, especially valuable to a rock analyst, have been made recently in the ultra-soft X-ray region, making possible the analysis of such light elements as sodium, fluorine, oxygen, nitrogen, carbon, and even the detection of boron. This has been achieved by the introduction of aluminum and other light-element X-ray targets in demountable X-ray tubes. Thin counter windows have considerably improved the intensities obtainable from such elements as magnesium, aluminum, silicon, phosphorus, and sulfur. Advances in manufacture of large *d*-spacing crystals have also helped the X-ray spectroscopist in this region.

X-rays behave like visible light. They can be refracted and diffracted, and they show interference. Their velocity is that of light. X-rays are generated when accelerated electrons collide with atoms. Because of their high energy, X-rays can in turn penetrate the outer electron shells and knock out electrons. When electrons from other shells fill the vacated positions, specific X-ray quanta of energy are emitted.

X-ray tubes have a filament and a metal target (Fig.38). Electrons emitted from the hot filament and accelerated by the voltage potential impinge on a positively charged target (anode), producing X-rays with wavelengths *typical* of the target metal by knocking out K, L, or M electrons. There is also produced considerable heat and a *continuum* or "white" radiation which is due to deceleration of electrons which have not collided with the inner shell electrons of the atom. The primary X-rays can produce secondary X-rays (often called *fluorescent* radiation), when they have enough energy to dislodge an inner shell electron. The fluorescent X-rays have the same properties as the primary X-rays when they interact with an atom. This secondary radiation is generally used in X-ray spectrography, despite the fact that it is weaker than the primary radiation, and mainly because it permits the use of sealed X-ray tubes as excitation sources. One advantage in using fluorescent X-rays is in the absence of the continuum.

The intensity of the continuous spectrum emitted from the X-ray tube depends on the target element. The maximum energy or minimum wavelength of this radiation depends on the potential applied. The maximum energy of X-rays emitted from the metals of a target is thus equal for all elements when the same operating

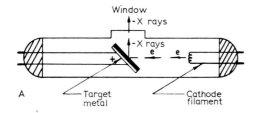

A

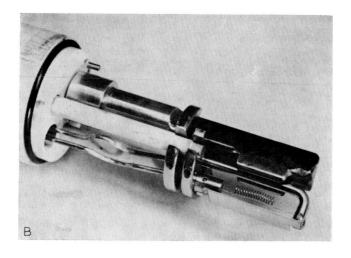

B

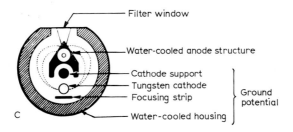

C

Fig.38. Schematic of a vacuum X-ray tube. A. Electrons from the hot filament travel in the direction of the anode or target where they collide with the target-metal atoms, producing X-rays which are permitted to pass through a thin window. B and C. The Henke aluminum target demountable tube and cross-section for the analysis of light elements. (Courtesy of B. L. Henke, Pomona College.)

voltage is used. It is important for the beginning X-ray analyst to keep in mind that while a certain high voltage will produce white X-ray radiation in a tube, the characteristic X-radiation from the target can only be excited when the voltage is high enough to give the electrons sufficient energy to knock out the K, L, or M electrons from the atoms of the target. Accordingly, we speak of K, L, or M absorption energy. When the *characteristic X-rays* from a target are the main source of secondary excitation, and because K-lines of elements with higher absorption edges (higher atomic number) than the target material can not be efficiently excited by this radiation, it is futile to endeavor to detect K-lines of elements of a higher atomic number than that used at the anode.

Since there are only two electrons in the K-shell, and not more than eight electrons in the L-shell, the X-ray spectra consist of only a few strong lines, characteristic of each element irradiated. This is the main advantage of the X-ray spectropgraphic method as compared with optical emission spectrography. In practice, for each element one need seldom consider more than three to four lines, and interferences are relatively few. Another advantage is the relative independence of these lines from the oxidation state or structural position of the atom analyzed (except for the very light elements). This is to be expected, because only the inner electron shells, which do not participate in chemical reactions, are affected. These properties permit nondestructive analysis of complex mixtures.

To disperse or separate X-ray radiations caused by different elements in the analyzed material, *analyzing crystals* are used. Here the principle of *diffraction* is utilized. Bragg has found that when X-rays of a certain wavelength, λ, pass through a crystal with the distance, d, between two planes of atoms, interference or reinforcement will occur at a certain angle of incidence, θ. This law is expressed in the fundamental equation used in X-ray spectrography as well as in diffraction:

$$n\lambda = 2d \sin \theta \qquad (19)$$

n is here some integer, indicating that various orders of reflection occur only at specific θ-values. From this it is clear that a *goniometer* or a device to change the position of the crystal with respect to the incident X-ray beam is necessary.

In addition to the analyzing crystal and the goniometer, electronic discrimination or *pulse height analysis*, P.H.A., is used to help separate characteristic lines which are detected at the same *Bragg angle* but are of different order (*n*, eq.19).

These are devices which can be set to accept only pulses with a certain amplitude distribution for counting, cancelling all other pulses (p.216). Fig.39 shows the essential parts for X-ray emission analysis and their relative position in an *X-ray spectrograph*.

After a pulse produced by an X-ray photon has been recorded it must be amplified. This is accomplished by a combination of a pre-amplifier placed close to the counter in order not to lose the faint signal through long connecting wires, and a linear amplifier for electronic processing of the pulses or counts. Amplified pulses are

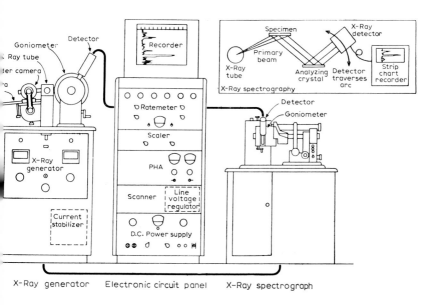

Fig.39. Norelco X-ray vacuum spectrograph and diffractometer. System with a diagram in right upper corner to show the conditions for X-ray spectrography. (Courtesy of Philips Electronic Instruments.)

usually sorted in an electronic discriminator after which they can be counted and recorded in digital form. Fast electronic counting and preferably printed output are essential.

Equipment needed to perform satisfactory analysis of elements from atomic number 12 (Mg) to atomic number 73 (Ta) consists of: a regulated power supply delivering 50–60 kV at 35 mA; one tungsten-target and one chromium-target X-ray tube for generation of hard and soft X-rays, respectively; a goniometer, connected to a vacuum chamber for the analysis of light elements; a set of analyzing crystals with $2d$-spacings from about 10 to 4 Å; a flow proportional counter for the light elements, and a scintillation counter for the heavier elements, and finally an electronic system similar to the one shown in Fig.39, to process and record the results. To effectively excite K-lines of elements above Ta, a 100 kV generator is necessary.

When light elements from atomic number 5 (B) to atomic number 12 (Mg) are to be analyzed, special targets such as the aluminum target demountable tube developed by Henke (Fig.38), and/or ultra thin X-ray tube and detector windows are also used. This is necessary to transmit and detect enough of the ultra-soft X-ray quanta which are greatly absorbed by all matter including air. Special

TABLE VIII

COMMON ANALYZER CRYSTALS FOR X-RAY EMISSION

Crystal	2d-spacing (Å)	Efficiency range (K-lines)	Efficiency range (L-lines)	Chemical composition
Lithium fluoride	4.03	K–Ce	Sn–U	LiF
Ethylene diamine ditartrate, EDT (or EDDT)	8.80	Si(Al)–Fe	Zr–Ce	$C_6H_{14}N_2O_6$
Pentaerythritol, P.E.T.	8.74	Al–Cl	Br–Sn	$C(CH_2OH)_4$
A.D.P.	10.64	Mg–Al (-Ti)	—	$NH_4H_2PO_4$
Gypsum	15.18	Na–Si (-K)	—	$CaSO_4 \cdot 2H_2O$
K.A.P., K-acid-phthalate	26.60	O–Si (-S)	—	$KHC_8H_4O_4$
Stearate	~100	B–O	—	$X(C_{18}H_{35}O_2)_2$ (X = Pb, Ba, Sr, etc.)

crystals such as gypsum and molecular films made of lead- and other stearates are also required to diffract the long-wave radiation. Table VIII lists the crystals and their characteristics used most frequently in X-ray emission.

Precision in X-ray emission analysis

Statistical reliability of analyses performed by counting pulses of gas-counters, scintillation or semiconductor detectors depends on several variables. The reliability of data depends in addition on two basically different factors, the *precision* and the *accuracy*. We have seen (p.36) that in classical chemical methods which are assumed to be inherently accurate (absolute) in skilled hands, the precision can be considered to indicate accuracy to a certain extent. In instrumental methods, however, the precision of counting does not necessarily result in a "true" value. The analyst here is recording relative intensities which can be influenced by so many factors that the reliability of the determination depends overwhelmingly on the ability of the worker to understand, control, and avoid a multitude of interferences. In actual work, this means that an instrumental analyst must know and guard against the elemental interferences, that may include mass absorption and spectral line effects, compositional variations as well as the physical differences in the samples to be analyzed. Stability and electronic drifts in the equipment, and even sample positioning and grain size effects are others. In short, the *accuracy* of the determination in instrumental analysis is greatly dependent on the skill of the analyst and his ability to select proper standards.

Precision of electronic instruments, to the contrary, is inherently very high, and mostly independent of the worker. This instrumental precision is a built-in feature of the counters and rate meters and it is directly proportional to the number of pulses counted. When we collect enough X-ray or gamma-ray pulses and plot them in the form of a histogram (p.34) and draw a frequency distribution curve, we see that it shows characteristics of the *normal* or *Gaussian distribution*.

We recall that the standard deviation was expressed on p.32 as the square root of the sum of squared single deviations from the mean, divided by the number of determinations minus one,

$$S = \sqrt{\Sigma \, d^2/(n - 1)} \qquad (20)$$

In this formula, suitable for so-called chemical statistics, the statistician seeks by mathematical means to give exceptional weight to the single chemical determination which is assumed to be relatively accurate as resulting from an "absolute" method.

When counts produced by pulses caused by X-ray quanta are recorded, it may be shown that standard deviation or *counting error* is:

$$\sigma = \sqrt{N} \qquad (21)$$

where N is the number of pulses recorded. The *relative counting error* or the *coefficient of variation* is:

$$\sigma \% = 100 \frac{\sqrt{N}}{N} \qquad (22)$$

Obviously, we may consider each count recorded in an instrumental method to be a representative of a single determination, and thus substitute the total number of counts recorded for N. Now we speak of *counting statistics* in contrast to the "chemical statistics".

In order to compare the precision of instrumental analytical results, the data should be based on the same number of counts, otherwise some of the compared values will be of higher precision than others.

The counting precision is easy to calculate and it must be selected according to the confidence limits desired. In Table IX, relative standard deviations are given to help the reader to preselect the precision desired by selecting the total amount of counts to be collected during each determination.

It can be demonstrated that in ideal cases when the precision is mainly dependent on counting statistics it does not matter whether one estimates the chemical percentages from the plotted working or calibration curves and uses these to calculate the "chemical sta-

TABLE IX

COEFFICIENTS OF VARIATION FOR A GAUSSIAN DISTRIBUTION

Counts collected	~67-% conf. limit (σ%)	95-% conf. limit (2σ%)	99-% conf. limit (3σ%)	Stand. dev. (σ) counts
100	10	20	30	10
1,000	3.2	6.3	9.5	32
10,000	1.0	2.0	3.0	100
20,000	0.70	1.4	2.1	140
50,000	0.45	0.90	1.3	225
100,000	0.32	0.63	0.95	320
400,000	0.16	0.32	0.49	640
1,000,000	0.10	0.20	0.30	1,000
5,000,000	0.05	0.09	0.13	2.500
10,000,000	0.03	0.06	0.10	3,000

TABLE X

X-RAY ANALYSIS OF CaO IN TWO ROCKS: COMPARISON OF CHEMICAL STATISTICS WITH COUNTING STATISTICS

Rock	CaO%	Chemical precision	Counting precision
Diabase	10.94	$n = 10$	$N = 409,600$ counts
standard	10.91	$\bar{x} = 10.939$	
W-1	10.94	$S = 0.014$	$\sigma = 655$ counts
	10.93	$C = 0.13$	$\sigma\% = 0.16$
	10.96		
	10.94		
	10.94		
	10.93		
	10.96		
	10.94		
Andesite	7.14	$n = 8$	$N = 409,600$ counts
	7.16	$\bar{x} = 7.145$	
	7.16	$S = 0.017$	$\sigma = 655$ counts
	7.15	$C = 0.24$	$\sigma\% = 0.16$
	7.15		
	7.16		
	7.12		
	7.12		

tistics", or simply bases the evaluation on the total number of counts recorded for one single determination. This is shown in Table X where data on carefully calibrated calcium oxide determinations in two rocks have been used to calculate the coefficients of variation (C and $\sigma\%$) by the two methods.

From Table X we see that chemical precision is comparable to the expected precision based on counting statistics only.

In actual unbiased routine, an analyst will mostly find that the predicted precision will tend to be better than the precision based on the actual chemical data. This is due to the dependence of most analytical methods on several variables which contribute to the total error. The standard deviation of a procedure is a combination of individual errors, each due to a different type of action, because a method consists usually of a system of such actions.

By definition (p.32) the variance is equal to the squared standard deviation, therefore, substituting standard deviation for variance in eq.1 (p.36) we can write:

$$\sigma = \sqrt{(\sigma_1^2 + \sigma_2^2 + \ldots + \sigma_n^2)} \tag{23}$$

This equation expresses the additive nature of variance, helping us to distinguish and separate a number of different factors influencing the standard deviation of the whole method, assuming that the experimental design is set up so as to permit the separation of factors of interest.

An attentive reader will have noticed that in the first example given in Table X, the "chemical" precision is better than the theoretical. This does occasionally happen in actual routine work. In such cases, if the difference is statistically significant, one may have some systematic errors which tend to cancel mutually. The tendency of an anlyst to discard certain "bad" results without a specific reason must also be suspect in such cases.

In the discussion of counting error above, it was assumed that all counts are due to the photon quanta representing the element of interest. Strictly speaking, this is never so because electronic equipment always generates a certain noise. In addition, the increment of the electromagnetic spectrum used as characteristic line always contains a certain percentage of photons which, although of corresponding energy, have not been caused by the excitation of the atoms of the element that is being analyzed in the sample. There may be interference from coinciding and neighboring spectral lines, from higher-order radiation, from scattered radiation, and from radiation from parts of the equipment containing the same element, which parts are exposed to X-rays due to poor collimation. Signals caused by fluctuation of the electric current, and even purely statistical events in the counter also interfere. All this interference is summed in the term *background* or *noise*. Obviously the true counting error will be influenced by the fraction of counts recorded which do not represent the element being determined. For example, approximating, one can state that if 50% of 100,000 counts recorded does not originate from the element of interest the relative counting error of the determination based on $5 \cdot 10^4$ counts will be 0.45%, in addition there will be an error of 0.32% based on the total number of counts collected (10^5 counts). The net error in this case is then 0.78%. Consequently, in precise routine work one has to count longer in order to register sufficient counts for a preselected precision when the contribution of the background becomes considerable. Working with major elements, one can ignore the background effect as long as the *signal to noise ratio* (or peak to background ratio: P/B, where P = total number of pulses counted, and B = pulses due to background) is higher than 10 and as long as at least 16,000 counts are collected. This gives generally a relative counting error of less than 1% ($\sigma\% < 1\%$). When the counting rate is low or when a smaller total count is used as a basis of analysis, and especially when the signal/noise ratio falls to figures like 5, 3, or 1.5, it is important to determine this effect in order not to report or expect better precision than theoretically possible. For this purpose, graphs such as in Fig.40 are used. It is obvious that the determination of the background becomes more important when the intensity of the signal decreases, as always is the case in *trace element* work. Here we can imagine the extreme case when no pulses can be identified as coming from the element sought and all signals counted represent the background. If these are accepted and reported on

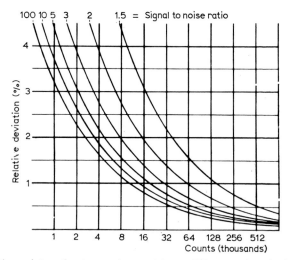

Fig.40. A graph to estimate counting precision at different peak-to-background ratios. (Courtesy of A. K. Baird, D. B. McIntyre, and E. E. Welday, Pomona College.)

different samples of different composition and with consequently different backgrounds, we would be measuring the range of the backgrounds of the different samples rather than the element, which occurs in this case in concentrations below the detection limits of the method. The precision thus reported would obviously have no relation to the element which actually has not even been detected. Therefore, in practical work with high background or with traces, it is very important to make standards and experimentally determine the lowest concentration of the element that produces a measurable increase in the spectral region of the characteristic line used, for example at the K_α-line. A weak peak in the proper 2-θ position or even an energy distribution maximum in the pulse height analyzer does not necessarily represent the radiation emitted by the element sought, and must be checked by addition of the element in careful work.

For the correction of precision for background we can use eq.23, remembering that individual counting errors can be expressed as square roots of counts recorded. Substituting the total counts recorded under the peak, N_p, and the number of counts representing the background, N_b, for the squared standard deviations, we get for the background-corrected standard deviation:

$$\sigma = \sqrt{(N_p + N_b)} \qquad (24)$$

To express this in relative standard deviation, which by definition denotes the precision in percent as related to the amount present, or the range of counts which represents this total amount in our case, we can write:

$$\sigma\% = 100\ \sqrt{(N_p + N_b)}/(N_p - N_b) \qquad (25)$$

This equation is useful especially in calculating precision in trace element analysis, when considerable background radiation is always present. Fig.40 is a graphic representation of this relationship useful in practical work.

METHODS IN X-RAY SPECTROCHEMISTRY

Most analytical methods in X-ray emission consist of irraditation of the sample with X-rays from the source and the measurement of emitted secondary radiation. Because the intensity of characteristic radiation is measured, all factors which may affect it have to be controlled. The method consists of irradiation of a fixed area of sample. The recording of pulses from the surface of this collimator-controlled area is performed over *fixed time intervals* or for time periods necessary to accumulate a preset number of counts (*fixed count*). Plotting of the counts per second versus percentages from different samples of known composition (standards) gives curves which permit the estimation of composition of the samples analyzed. In order to estimate the concentration of an element in an unknown sample, samples of similar composition with known concentration of this element have to be irradiated under precisely same conditions. Only when the "relative intensities" of *known samples* or standards are plotted against their composition in a coordinate net, is the analyst able to derive the approximate concentration of the

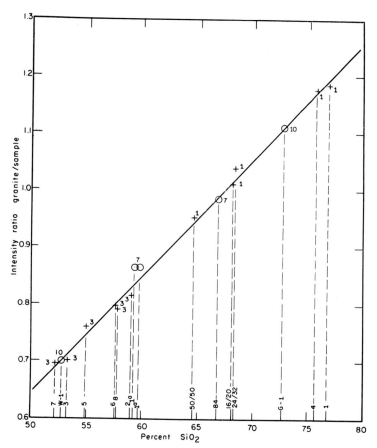

Fig.41. A working curve plotted for silica in different rocks varying in composition from basaltic to granitic. All intensity ratios are calculated in relation to a granite sample of approximately mean composition of the rocks analyzed, kept in the same position in the X-ray spectrometer during the determination. The calibration curve is drawn through the points representing standard rocks W-1 and G-1. As long as the "granite sample" remains intact, an unknown sample may be irradiated and the composition estimated by plotting it on the ordinate, seeking the point where a horizontal line drawn from it crosses the calibration curve, dropping a vertical line to the abscissa and reading the composition.

element sought in the unknown sample. This is done by plotting the relative intensity figure or composition of the unknown on the ordinate and projecting horizontally toward the calibration curve and reading the concentration on the abscissa as shown in Fig.41.

A very useful plotting method in geochemistry for major element analysis where no serious background problems exist is to calculate the ratio of the counts recorded from the standard and sample, and plot it on the ordinate instead of plotting counts/sec or counting time.

From the above, we can conclude that should the condition of the samples in some way affect the intensity, this method would not give accurate results. In addition to concentration, the main factors that control the intensity of the secondary X-ray radiation, assuming that a fixed area of a powdered sample is irradiated, are: (a) grain size; (b) thickness of sample; (c) nature of the surface; (d) physical nature and varying amounts of the (mineral) constituents; (e) varying amounts of spectrally interfering elements; (f) elemental mass absorption, and enhancement effects; (g) oxidation changes of the surface and contamination by adsorption of oils or "blossoming out" of water-soluble salts during storage.

The importance of these possible interferences in X-ray spectrography is demonstrated by the fact that, depending on the particular needs and analytical considerations, the methods used by different analysts tend to vary greatly, mainly in sample preparation, and arguments for mathematical computerized mass absorption corrections as opposed to arguments for built-in analytical routine corrections based on compositional similarity of samples, can be heard.

If we ignore sample preparation methods which lead to aqueous solutions and special methods for small samples or thin samples, we may state that the most straightforward sample preparation would be the use of *plain powdered samples*, spread, for example, over plastic films (mylar) in sample containers. When we analyze such

samples for major constituents, we soon discover, however, that, especially in the analysis of light major elements which represent the bulk of the earth's crust, the precision and intensity greatly suffer due to the difficulty in packing of loose powders to the same density, and due to the absorption of the soft radiation by the plastic film. To avoid these disadvantages, sample powders are usually compressed to pellets. For example, in the author's laboratory, the following method is used:

Procedure for pelletizing

(*a*) Use a fine powder with 95% of grains passing the 325-mesh screen (or a 400-mesh screen). This minimizes grain size effects which are considerable if powders of different grain size distribution are used.

Homogenize this powder if the sample has been stored for long, or transported, by tumbling in homogenizers (p.15). If sample has been just ground in vibrational type mills (Pica), the homogenization can be omitted. Do not sieve the sample for purposes other than to establish the time necessary for grinding. Sieved samples can become unrepresentative due to preferential grain size distribution of minerals in rocks.

(*b*) To provide *reproducible surfaces*, press the fine sample at about 30,000 p.s.i. (Fig.42) against a hard, flat, chrome-plated steel, or glass surface, preferably in a specially hardened-steel die (Fig.43) designed to produce a pellet with plastic frame and backing (Fig.44).

(*c*) To assure uniform sample thickness, spread one gram (1 g $\pm$ 10 mg) of rock or mineral powder over an area of about 6 cm^2 (6 cm^2 $\simeq$ 1 sq. inch). Prepress by hand with the inner sleeve, using the long piston (Fig.43). Do not shake to spread the powder, which may segregate mineral particles. One gram of powder pressed corresponds to about 1 mm thick layer. This provides an *infinite thickness*

Fig.42. Soiltest press used in the Nevada Mining Analytical Laboratory for making pellets of powdered rocks. In foreground the pressure die used. All sample preparation is done in a "geochemical" preparation room.

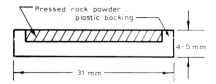

Fig.44. Cross-section of a pressed rock-powder amberlite-backed pellet.

of sample for elements up to Zr, insuring that all K_a X-ray quanta
($< Zr$) are absorbed within the sample and thus generate the
maximum of secondary radiation, which in turn can penetrate back
to the surface to be counted. Pellets with thicker rock-powder layer
have been found less durable. Therefore, close sample weight control may be selected as an alternative to assure constant intensity
from heavier elements in trace element work.

If the pellets crack up upon removal from the die, which is often
the case with tuffs, add 2–3 drops of distilled water to the fine-
powdered sample before final pressing. Remove the inner sleeve
carefully and cover the prepressed rock powder with Amber
Bakelite 1386AB powder manufactured by Buehler Ltd. This resin
has been shown to have good storing and pressing properties when
compressed with rock powders. Boric acid has also been found suitable as pellet backing. The amount used is roughly measured to
achieve constant pellet thickness. Press at 30,000 p/sq. inch, wait
15 sec, release the pressure carefully and remove the pellet from the
sleeve using the cup in the lower left corner of Fig.43.

Pellets obtained in this fashion have been found to be durable for

---

Fig.43. Pressure die designed by Baird, to prepare plastic-backed pellets
(second piece, lower left corner) of rock powders for X-ray spectrographic
analysis. To obtain smooth, contamination-free surfaces, the N.M.A.L. uses a
glass disc placed under the sample before pressing. The glass breaks but the high
pressure prevents it from disintegrating prior to the removal of the pellet. The
long piston (upper left corner) was also added in this laboratory to prepress
the sample powder, thus assuring equal thickness. One should avoid shaking the
powder to prevent segregation and the pistons should not be twisted, which
causes metal to rub off the walls. In upper right corner, a screw-top container,
"rabbit", is shown. In fast-neutron activation (p.316), powdered rock samples
are packed ino these vials. The scale for the pressure die is in inches, for the
"rabbit" it is in cm.

years if stored in a dry room. A minimum of cracking or crumbling has been observed. However, before putting these pellets into the vacuum of the spectrograph, it is advisable to keep them in vacuum in a desiccator for about two hours. This prevents occasional contamination of the X-ray tube if the sample is placed in an inverted position above the source as in the Philips vacuum X-ray unit.

The procedure above equalizes grain size effects, assures infinite thickness of sample for the major light elements and a constant thickness for heavier elements, and a flat reproducible surface. It also produces pellets of comparable density. Naturally, pelletizing can be applied to either plain rock powders or to reground fluxed material.

Nondestructive X-ray emission analysis procedure

Place the powdered rock pellets directly into the X-ray spectrograph, select the desired 2-θ position, and record relative intensities compared with pellets made by the same pelletizing process of similar standard materials. Read in the sequence standard–sample–standard . . . , repeating four times. Take the means of the four determinations of standard and sample and calculate their ratio assuming a linear relationship. Multiplication of the concentration value of the standard by this ratio will give the concentration of the element in the unknown. Take another standard, preferably selected so as to bracket the unknown, and repeat the procedure. If the sample is closely bracketed, a mean of these two results will usually give a more accurate figure. If the "unknown" sample is known to be chemically and physically more similar to one standard than to the other, then the ratio of the more comparable pair is used, and the second standard is used only to indicate the approximate slope of the calibration curve in this region.

In spectrographs with multiple sample holders, one places the two standards in the first and last positions, respectively, with the unknowns spaced between, and irradiates, for example, in the sequence: standard A —sample—sample—standard B. Taking the *intensity ratios* of analyzed pellets instead of absolute counts/sec eliminates the need to rigorously control the electronic stability of the equipment and minimizes the effect of electronic drifts common to all equipment of this type. To further minimize short period drift, it is advantageous to count in a zig-zag manner, for example, in the sequence 1–2–3–4–4–3–2–1. In computerized X-ray spectrographs with eight sample holders, one can analyze three individual two-pellet samples spaced between two bracketing standards in this sequence. A high degree of reliability and virtual freedom of possible drift effects is thus achieved.

Discussion of the method

This method appears to give linear curves for silica (Fig.41) over a composition range covering common silicate rocks. Similar curves are obtained for other major light elements.

This is a truly *nondestructive technique* where the sample is neither fluxed nor mixed nor diluted with any binding or absorbing material. In terms of sample preparation, this is the most straightforward approach because it does not require weighing of the sample. For light elements approximately 1 g of powder can be taken by simply marking, for example, a test tube and using it as a scoop for the powder. To assure a similar total thickness of pellet, the binding or backing powder is also taken with a marked scoop.

The major *advantages* of this method may be in applications where rapid, precise, and reliable data are required in the analysis of physicochemically similar samples and where good chemically analyzed standards of similar composition are available, and in *geochemical studies*, where one usually compares compositional variations of similar rocks within the same rock body or pluton. In geochemistry, one is often more concerned with the relative variations in composition and the ability to detect with certainty that these variations exist, rather than with absolute values as long as the approximate range is known. Excellent and useful practical work has been performed even with optical semiquantitative methods where the precision is often at least about one order of magnitude less than that of the X-ray emission methods.

In *industry* one also is mostly concerned with the quality control of raw materials and products which are purposely selected for minimum compositional variations and standardized uniform physical characteristics. A certain type of rock, ore, cement, ceramic, glass, slag, a metal or an alloy, is usually analyzed in industry to detect relatively small differences. Detection of compositional variations to pinpoint deviations in processes or changes in raw material composition is the most common duty of an industrial chemist. Even the crudest form of nondestructive analysis from the standpoint of a laboratory chemist, the so-called "on-belt" X-ray emission analysis, has found application in these industrial fields because of the all-important time factor involved. X-ray machinery has been specifically built for automatic "on-belt" analysis.

Speed is often a major factor, for example, a mine pit foreman needs to know the composition of the rock while it is being moved in order to make a decision where to blast next, and an industrial product's composition often has to be known while it is manufactured in order to avoid costly production of large quantities of substandard material.

In general, simplicity and speed of analysis often result in advantages which may weigh more than the better accuracy achieved by meticulous but time consuming analysis. An analyst must always endeavor to select the most straightforward method that will produce the desired reliability. Theoretically, the results obtained by instrumental methods can be no more reliable than the absolute data on which these results must always be based. If the gravimetric analysis of a standard sample has resulted in a figure which is actually off by 1% from the "true" value, no quantity of instrumental determinations based on this value, no matter how precise, can give the "true" value except by a purely statistical chance due merely to the random distribution of the relative readings around the assumed true value. And even when this happens, one has no means to detect this event at this stage of analysis (see p.8). Thus, an instrumental analysis can only be as good as the standards used. Therefore, a practical analyst's question should be: if I use the standards or reference samples available and base my instrumental procedure on these, will I get results which are acceptable within the customary limits of the absolute or basic method by which the standard samples were analyzed? If the analyst can prove that this is the case with his method, he has shown that the method produces results acceptable for standards used, as well as for the unknown samples. He can go no further in terms of accuracy, but he can certainly go farther in terms of precision alone, and this is where the great advantages of the high precision of instrumental methods are.

For example, let us assume that by some well established gravimetric procedure we are reasonably certain of being able to achieve a reliability of 1%, and we have analyzed two samples, each showing a concentration of 10% of an element X. If we want to work with a 99% confidence, which means roughly a range of 6σ or 6% (actually 5.2σ), we can assume that each of these samples has a composition within the range of $10 \pm 0.3\%$, or a total range of 0.6%. We can say no more at this stage, no matter how many analyses we may perform, because 99% of all results will group symmetrically around this value and within these limits. If, however, we decide to *compare* these samples using an instrumental method which is sensitive enough to detect one part in a thousand, and which gives us a precision of 0.1%, and upon excitation and measurement of intensities from both samples we get 998,000 and 1,001,000 counts, respectively, for these samples, we can say, if we have numbered the samples, that sample no.1 has approximately 0.3% of the element less than sample no.2. Thus, comparing the relative compositions (intensities) of two samples, we get additional valuable information about their nature. This information permits us to distinguish between the individual samples of a given composition range. If enough similar samples falling within the same range are analyzed, and we find that their spread covers the total range of composition, we may be able to find a mean value $(\bar{x})$ and report the data in absolute units, again assuming that the absolute methods have yielded accurate results. Data given in instrumental trace element analysis can usually be traced back to absolute methods in similar fashion and are thus not absolute in a strict sense. It has been emphasized before that the precision necessary to do good trace element work is orders of magnitude less than the precision required in major element work (p.52). Only continuous accumulation and evaluation of data coming from different laboratories and retrieved by different methods finally form the standard basis of elemental analysis in major element as well as in trace element work.

Coming back to the evaluation of the usefullness of the most straight-forward nondestructive methods we can readily see that in geochemical work we may use these procedures to detect minor compositional differences in similar rocks. To a certain degree, we can consider *silicate rocks* as *dilutions* of cations, such as Al, Fe, Ca, Mg, K, and Na in a stable silicon–oxygen tetrahedron framework. The amount of silica (SiO_2) varies in the silicate rocks usually between 40 and 80 wt.% and the major cations occur in dilution ratios from approximately 1/100–1/5. Similar considerations can be applied to carbonates, phosphates, and sulfates, and to a certain degree to oxides as well. All these anions represent ready-made dilutions in relatively light matrixes, which considerably decreases the absorption and enhancement effects (p.265). Therefore, *further dilution* of silicate, carbonate, phosphate, or sulfate powders, unless performed for special purposes, should not greatly benefit the nondestructive X-ray emission analysis of these substances. In contrast, the intensities would be affected, especially in light elements which would mean lower counting rates and less precision unless longer counting periods were used. Especially in trace element analysis, dilution may mean poorer detection limits by one order of magnitude, which in the case of X-ray emission may mean practical detection limits of 100 p.p.m. instead of better than 10 p.p.m. in solid sample.

By avoiding the mixing of samples with fluxing and heavy absorbing media when this is not necessary (see p.265), one avoids trace element contamination. In rocks rich in crystalline, zeolitic, or other essential water or volatile constituents, fluxing can change the relative composition of sample. Unless this is accounted for (p. 267), and standard has been so selected as to contain a similar amount of the volatile component (which would be very cumbersome in routine work), the results would be inaccurate. This, because relative intensities affected by bulk compositional changes are measured and not absolute intensities. For example, a sample containing 10% water lost upon fluxing with a dry flux will be concentrated in proportion to this loss. Unless weights are balanced this sample can not be compared to another sample not containing a similar amount of water, even if fluxed and treated similarly. Since many sulfates, phosphates, volcanic rocks in general, tuffs, perlite, and obsidian contain considerable water which is not lost in the vacuum of the X-ray spectrograph, the example given must be considered real. In general, any kind of radiation background introduced will lower the detection limits, especially of the trace elements, regardless whether it it due to scatter or the L-spectrum of the heavy absorber or to elemental trace impurities.

The *disadvantages* of the truly nondestructive method are in the necessity of bracketing the unknown sample closely by *chemically* and *physically* similar substances. The reason for this is that the intensity of X-ray emission from an element is considerably affected by the *mass absorption coefficients* of all the elements in the mixture, thus depending on relative concentration of elements present.

If we consider that each element has a specific density which has a specific absorbing effect on X-radiation, we can easily see that the total mass absorption coefficient of a mixture for a given element will depend on the elemental composition of this mixture. In the simplest case when the sample is composed of a 100% pure element M, the intensity, I, of the fluorescent radiation will be dependent on the total amount of atoms irradiated in a monoatomic layer. The intensity will increase when additional atom layers are added till the *infinite thickness* for this element is reached.

This maximum intensity, I_{max}, at infinite thickness will depend on the mass absorption coefficient characteristic for this element. In an ideal case, we can assume that at infinite thickness of a composite sample a weight-fraction of element, M, would produce characteristic intensity, I_M, in direct proportion to the percentage of this element, M%, in the mixture.

$$I_M \text{ (sample)} = I_{max} \cdot M\% \qquad (26)$$

This would be true if the mass absorption coefficients for other elements in the mixture were equal, but in relation to element M, lighter elements absorb radiation less and heavier elements more. Therefore, if lower atomic number elements are present, relatively more characteristic radiation of element M will emerge from the sample than that which is directly proportional to its weight fraction. This will cause a *positive mass absorption effect*. If higher atomic number elements are present, the effect will be negative. Thus, strictly speaking, the total mass absorption effect on a characteristic X-radiation will depend on the composition of the sample and can be either positive or negative. If the lighter elements dominate, the mass absorption effect will tend to be positive, if the heavier elements form the main bulk, the effect will tend to be negative. A certain degree of cancelling of the opposite components will also occur in complex mixtures.

These factors result in nonlinearity of calibration curves in systems. For example, a heavy element in a light matrix will show an upward bulging of the analytical curve, and an element lighter than the matrix will tend to give a downward bulging curve, as shown in Fig.45.

In addition to this effect caused purely by density differences between the element sought and the matrix, we know that X-rays can excite X-rays of lower energy. This process happens to be most efficient when the wavelength of the impinging radiation is just above the absorption edge of the element. So that, for example, iron

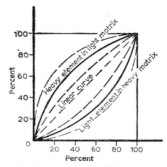

Fig.45. A family of possible X-ray calibration curves demonstrating convex (bulging) forms, resulting from plotting X-ray intensities versus content in percent of a heavy element in a light matrix. Concave forms result when a light element is determined in a heavy matrix.

in the sample will excite manganese much stronger than calcium or potassium, and copper or nickel in the sample will excite iron much stronger than titanium. This means that elements one or a few atomic numbers higher than the element analyzed, when present in the *matrix*, may excite this element. This excitation is in addition to the one caused by the X-ray radiation from the tube. In X-ray emission spectrography, this is called *enhancement effect*. Naturally, when the enhancing X-radiation from the matrix is absorbed by an element in the sample this causes a specific decrease of intensity of this radiation and thus a *negative* effect.

The *enhancement* and *absorption edge* effects (not to be confused here with mass absorption treated above) are due to proximity of X-ray energies emitted within the sample. This means that in complex mixtures one must especially watch for mutual interference of elements with close atomic numbers. When we analyze elements with close atomic numbers, densities are usually similar so that the mass absorption effect becomes less significant, whereas the enhancement and the absorption edge effects dominate and can override or contribute to the effects of different mass absorption coefficients. In either case, a nonlinear calibration curve will result unless these combined effects happen to cancel each other.

Before giving an example, the reader must be reminded that elemental densities do not necessarily increase with increasing atomic number. Density of an element depends on its atomic size as well as on structural packing in the pure state. Generally, but with exceptions, densities increase from the first (I) and second (II) main groups in the periodic system and through the third (III) and eighth (VIII) sub-groups with increasing atomic number, but generally decrease somewhat with increasing atomic number starting with the first (I) and second (II) sub-groups and proceeding through the third (III) and seventh (VII) (also the 0-group) main groups of this system.

We know that the addition of new electron shells increases the size of the atom and that the addition of valence-electrons to a completed shell has the opposite effect. The addition of non-valence electrons to an incomplete shell in the first, second, and third long periods, starting with Cu, Ag, and Au, respectively, does not affect the atomic size appreciably while an irregular small decrease can be noted (with exceptions). A remarkable similarity in atomic sizes in elements of the second and third long periods results from an effect called *lanthanide contraction*. This effect is partially caused by addition of electrons to an inner shell of the atom while the outer valence-electrons within the same period retain a similar configuration. This causes the atom to contract.

It is within this group of elements with structures mostly of the hexagonal closest packing type that we may best see the effect of the atomic size on the density and the dominating effect of structural packing on density and the atomic radius of the element. Table XI shows the interrelationship when we compare the increasing density with the decreasing atomic radius in the hexagonal closest packing (HCP) with the effect of the looser packing of the face-centered cubic (FCC) or the body-centered cubic (BCC) types.

Returning to the enhancement effect of X-rays emitted within the sample on elements standing close in the periodic system, we have the example of aluminum oxide, Al_2O_3, and silicon oxide, SiO_2, mixture, important in clay mineral studies. In this case, the silicon

TABLE XI

THE EFFECT OF PACKING AND ATOMIC RADIUS ON DENSITY

Atomic number	Element and symbol	Atomic radius	Packing type	Density
57	lanthanum, La	1.87	hexagonal closest packing	6.2
58	cerium, Ce	1.81	face centered cubic packing	6.7
59	praseodymium, Pr	1.82	hexagonal closest packing	6.8
60	neodymium, Nd	1.81	hexagonal closest packing	7.0
63	europium, Eu	1.99	body centered cubic packing	5.3
64	gadolinium, Gd	1.79	hexagonal closest packing	7.9
65	terbium, Tb	1.77	hexagonal closest packing	8.3
66	dysprosium, Dy	1.77	hexagonal closest packing	8.5
67	holmium, Ho	1.76	hexagonal closest packing	8.8
68	erbium, Er	1.75	hexagonal closest packing	9.0
69	thulium, Tm	1.74	hexagonal closest packing	9.3
70	ytterbium, Yb	1.92	face centered cubic packing	7.0
71	lutetium, Lu	1.74	hexagonal closest packing	9.8

K_a photons have just enough energy to excite the aluminum K_a radiation. In this process, the silicon radiation is strongly absorbed, whereas the aluminum radiation is enhanced. It so happens that the mass absorption coefficient is also lower for silicon than for aluminum, despite silicon having a higher atomic number than aluminum (Al has tighter FCC, whereas Si has looser diamond-type structure). This also would result in enhancement of the aluminum radiation in a silicon-rich matrix and the absorption of the silicon radiation in an aluminum-rich matrix. Both of these factors, the enhancement and the mass absorption, act in this case in the same direction producing strong non-linear effects for this mixture. This case is demonstrated in Fig.46.

It is clear from the above said that the interference due to differences in density as well as to the absorption edge effect overlaps and that the total effect may be positive or negative, causing in both cases nonlinearity in calibration curves.

If one now examines the curves in Fig.46 and recalls true composition ranges of silicon and aluminum in *igneous rocks*, where alumina, Al_2O_3, generally varies from 15–20%, and silica, SiO_2, from 40–80%, one may assume that the effect of a 5% range of aluminum on silicon would be negligible but that the effect of silicon on aluminum intensities could still be considerable. Should we now be able to determine experimentally these mutual effects

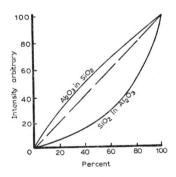

Fig.46. The system SiO_2-Al_2O_3 showing strong absorption and enhancement effects. (After of R. K. Scott.)

and should they turn out to be of a magnitude less than the reliability of the absolute methods, we may accept these errors in a routine method while introducing our approximate empirical corrections whenever possible. This brings us again to consider the applicability of the *truly nondestructive method.*

It is obvious from the discussion above that the proper corrections should account for the density and the composition as well as the absorption edge effects on the X-ray intensity. All these effects have simple additive characteristics but, to perform a complete correction, considerable calculations are necessary (p.272–277). What is even more discouraging to a practical analyst is that to correct for one element, all other main constituents of the sample must be known.

To avoid these cumbersome calculations, analysts have attempted to develop methods which would minimize these effects as much as possible. Before presenting an accurate method with mathematical absorption and enhancement corrections, *dilution*, *heavy absorber*, and plain *fluxing* methods have to be discussed first.

When the analyst evaluates or compares methods he must first ascertain the purpose and range of applicability of each method under study. Much misunderstanding and confusion can result when one attempts to compare methods incompatible and designed originally for different purposes, or methods covering wide ranges of composition with methods worked out for narrow concentration ranges. When "relative intensity" or instrumental methods are evaluated, the above statements gain more importance than if absolute methods are employed. This is because the whole philosophy of instrumental methods is based on relative rather than on absolute values and such an analysis is always a compromise to a permissible degree. An analyst needs a method only as reliable and fast as the required precision and accuracy of the results requested will permit. Nothing is gained by spending more time and effort to get results better than necessary to solve a specific problem, unless the problem is to investigate precision and limits of applicability of a method.

Thus, we must start by clearly separating two analytically different problems. One of these is: (*a*) to analyze by X-rays the widest possible composition ranges of mixtures, rocks, minerals, and artificial products, for which it is difficult to find enough good standards; the other, (b) to analyze common materials of restricted composition for which good standards exist.

In the first case, if we work with rocks, minerals, ores, oxides, sulfates, carbonates, or phosphates, for which we want to prepare a good working curve without being bothered by absorption and enhancement effects, and also if we want to avoid the physical effects due to different crystalline mineral phases and heterogeneity effects due to different preferred grain shapes (mica versus feldspar) and their preferred orientation, we will seek to: (a) flux the sample; (b) dilute the sample; and (c) add heavy absorbing elements in some cases.

Fluxing homogenizes the sample and results in a glass where mineralogical effects disappear. Sodium and lithium tetraborate, or lithium carbonate, are good fluxes for this purpose. *Dilution*, which is accomplished if a fluxing agent is added can be of varying degrees depending on concentration and effect desired. It diminishes the mass absorption effects and the enhancement effects. *Addition of heavy absorbers*, such as barium or lanthanum oxides (BaO, La_2O_3) to the flux, increases the mass absorption to such a degree that the individual variations of sample composition become relatively insignificant and the mass absorption effect, as well as the enhancement effects, decrease. Naturally, this procedure also decreases the overall intensities.

A combined procedure incorporating fluxing, mild dilution, and the addition of a heavy absorber has been developed by Rose et al. (1963) at the United States Geological Survey. This procedure gives good results over a wide variety of compositions. It is given below in a somewhat modified form as an example of a convenient overall method of X-ray emission analysis.

It is especially applicable in laboratories testing the composition of rocks, minerals, ores, concentrates, unusual samples, prospector's samples, raw materials, and for the identification of unknown samples and analyzing for their preliminary composition in the qualitative as well as in the quantitative sense.

Heavy absorber procedure by Rose, Adler, and Flanagan

Mix 250 mg of the −325-mesh sample powder (p.16) with 2,000 mg of dry $Li_2B_4O_7$ powder, and 250 mg of dry La_2O_3 powder in an agate or mullite mortar. Transfer all of this mixture into a 30–40 cm³ graphite crucible. First fuse carefully at about 750°C until raising and bubbling, due to escaping H_2O and CO_2 gases, ceases. Raise the temperature to full blast of the burner or place the crucible into an oven maintained at about 1,000°C. Fuse an additional 10 min. Cool and weigh the button. This is done by either removing the button and weighing it separately, or by weighing the crucible with the button, then removing the button, and taring the crucible. Add sufficient lithium tetraborate to bring the combined weight of the flux to 2,600 mg, and flux again. Cool and crush the button in a cylindrical hardened-steel mortar without grinding, transfer all crushed material into a small ceramic or hardened-steel vibration-type ball-mill and grind for a predetermined time (3–6 min) to 95% < 325-mesh grain size. Pelletize as shown on p.251. If four instead of one or two pellets are desired, double the amounts of sample, flux, and absorber.

This method gives linear curves over a wide range of composition for silica. Nevertheless, comparison with standards of similar composition is recommended.

Lithium-tetraborate fusion procedure by Welday, Baird, McIntyre, and Madlem

Welday and co-workers from Pomona College have evaluated the reliability of different methods of sample preparation in X-ray emission analysis. The special purpose of this work was to find a method suitable for *light element* analysis. We must recall here that seven out of eight major elements of the earth's crust—O, Na, Mg, Al, Si, (K), (Ca)—can be considered light elements. Based on a computerized statistical study on 68 individual granitic rock samples, these authors have selected a moderate dilution-fusion with lithium tetraborate, without the addition of a heavy absorber. Welday's method, slightly modified, is given below.

Samples suspected of containing considerable *volatiles*, H_2O, and CO_2:

Dry 2800 $\pm$ 5 mg of sample powder of grain size under 325 mesh, spread over a tared porcelain or platinum crucible in oven (preferably vacuum) at 105°C for 2 h. Determine the weight loss ($-H_2O$). Heat at 850°C for 10 min in a muffle furnace, cool in a dessicator (vacuum preferred), reweigh and determine the fusion loss. *Total fusion loss* = ($-H_2O$) + fusion loss at 850°C. Oxidizing substances may cause weight gain.

All samples, whether the fusion loss has or has not been determined:

Tar a graphite crucible of 30–40 cm³ capacity, preferably using an automatic top-loading balance (Mettler P120). Add 5,200 $\pm$ 10 mg $Li_2B_4O_7$, spectrographic grade, maximum moisture 0.4%. Add 2,800 $\pm$ 10 mg of specimen powder (unfused). Mix with a spatula as well as possible, trying not to scratch the soft crucible walls. Fuse for 20 min at about 1,100°C in a muffle furnace. If necessary, preheat for 10 min at 850° in another furnace to avoid splattering.

Cool, and reweigh the bead, which should be 8.00 g, less the fusion loss determined, if any. A significantly larger weight difference may indicate a weighing error. Crush the bead in cylindrical hardened-steel mortar, avoiding twisting action, and grind the chips for 3–6 min in a vibration-type mill to pass a 325-mesh screen. Press as shown on pp.250–253 at 30,000 p.s.i. The handling of several samples simultaneously does considerably reduce time required for weighing.

This method gives higher counting rates than the heavy absorber method and, although shown to be applicable with granites and schists, it may prove to be applicable for other rocks as well. Obviously this method can not cover as wide composition ranges as those given for silica in the heavy absorber fluxing method described above. Both methods have definite advantages over the plain pressed rock-powder method when micaceous rocks, especially *biotite-bearing*, are analyzed, and if rock standards with similar amounts of biotite are not available.

Method of Norrish, Sweatman, and Hutton

The advantages of fluxing and dilution in a light matrix, and the benefits of a heavy absorber addition have been combined with *corrections for mass absorption* in a method developed by Norrish, Sweatman, and Hutton at the Commonwealth Scientific and Industrial Research Organization, Division of Soils, Adelaide, S.A. (Australia). This method is given in detail here because it represents perhaps one of the most thorough attempts at maximum reliability and is adaptable to computerized processing. The method is designed to handle samples of variable composition.

Preparation of flux. Mix 19.0 g of anhydrous lithium tetraborate, $Li_2B_4O_7$, X-ray grade; 14.8 g of lithium carbonate, Li_2CO_3, spectrographic or extra pure grade; and 6.60 g of pure lanthanum oxide, La_2O_3. (Prior to weighing, heat the borate and the carbonate powders at 500°C and the lanthanum oxide at 900°C if moisture is suspected.) Flux at 1,000°C for 20 min or till effervescence ceases in a large platinum or graphite dish under cover. If a large enough vessel is available, fuse double the quantity given. The volume does increase during the fusion.

Pour the melt over a thick ($\sim\frac{1}{2}$ cm) polished aluminum sheet, spreading it over a wide area to prevent fusion of the cooling glass with aluminum. After cooling, collect the shattered pieces, and crush. Store the mixed flux, well stoppered, in a dry place.

Preparation of pellets

Mix 1,500 $\pm$ 10 mg of this flux with 280 $\pm$ 1 mg of sample (sample is ignited prior to fusion and weight loss determined) and 20 mg of sodium nitrate, $NaNO_3$, in a graphite or gold plated platinum (Palau) crucible. (Wetting occurs if platinum crucible is used.) Use a platinum wire loop for mixing. Flux at 980°C for 10 min. (This temperature should not be exceeded when gold plated platinum crucibles are used in order to prevent their quick deterioration; check furnace temperature by placing some NaF, which melts at 980°C, into crucible.) Stir the melt with the same platinum wire before pouring, and reheat briefly if necessary to assure homogeneity of flux. Pour quickly onto a graphite disc preheated on electric plate at about 230°C. Pouring is done through a preheated short brass or aluminum cylinder with inner diameter large enough for the sample holder of the X-ray spectrograph ($\emptyset$ ∼1.25 inch, ∼32 mm, for Philips-Norelco sample holders). Press melt evenly into a glass pellet. Withdraw the plunger and remove the cylinder. Anneal the sample between clean compact asbestos mats for 10 min at 200°C, remove from electric plate and let cool between the mats. Smooth off projecting edges with file and store in a labeled envelope. Fig.47 shows the equipment required for this procedure.

Remarks. If powdered pressed samples are preferred, the melt can also be cooled in the crucible, crushed, and ground in a vibration mill, and pressed (Fig.48). The graphite crucibles and the disc can be machined from graphite rods to desired sizes. The graphite discs have a slightly concave upper surface. This tends to collect the melt drop in the center and spreads it evenly over the whole surface when pressed by the piston or plunger (Fig.47). Graphite tends to somewhat contaminate the surface of the pellets. This affects the intensity of light elements by depressing the Mg, Al, and Si counts by about 1–2%. This can be alleviated by placing the discs in cold water, but more pellets tend to break as a result of this treatment. Sulfide or sulfate sulfur is reduced by graphite and lost from samples. In gold plated platinum crucibles, sulfate sulfur is not lost. The upper surface of the glass pellets is not entirely smooth, but this does not seem to affect intensities nor precision as long as the surface is not bent. These pellets can be considered infinitely thick for wavelengths longer than Mo K_α, $\lambda = 0.71$ Å.

Correction factors for *mass absorption* and *enhancement* for *this specific flux* have been calculated for oxides and partially determined experimentally by the Australian team mentioned above. These factors are given in Table XII.

Procedure

To apply these correction factors, the following equation is used, using silicon as an example:

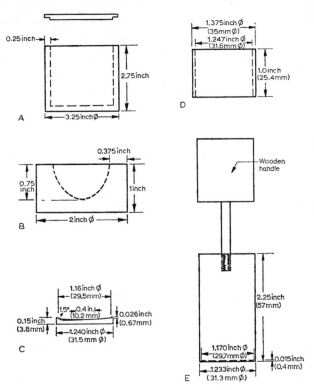

Fig.47. Drawings of graphite and other parts for preparation of fused pellets for X-ray spectrographic method of Norrish. A. Graphite dish. For preparation of mix. Made from graphite rod. A.G.X. grade. B. Graphite crucible. For use in muffle furnace. Made from graphite rod. A.G.X. grade. C. Graphite disc. D. Brass ring. E. Aluminum plunger. (Courtesy of K. Norrish, C.S.I.R.O., Australia.)

Fig.48. Pressure die drawings. The steel cylinder (A) is placed on the basal cylinder part (B). The aluminum tube (D) having an easy fit in the cylinder (A) is inserted into (A) and the sample powder (0.2–1.5 g) is poured into (D). The transparent plastic plunger (F) having an easy fit in (D) and being polished on both ends is then used to spread the sample evenly. The sample can be observed by looking through the polished end of the plunger (F). By pressing on the plunger—instead of the plastic plunger (F) a steel plunger made as (F), can be used—and by rotating the tube (D), and pressing on (D) and rotating the plun-

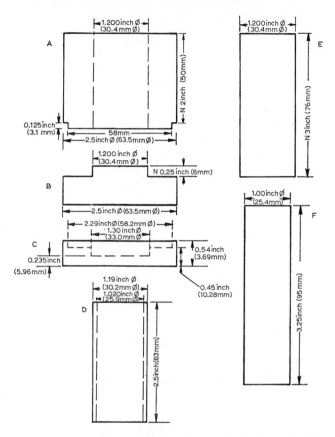

ger (F, steel or plastic) the sample is freed so that if the plunger and tube (D) are withdrawn carefully, the sample is left as a flat disc on the base part (B). A binder powder (amber bakelite or boric acid) is poured into (A) to encase the sample. The amount of binder is roughly estimated to achieve a final thickness of pellet of 4–5 mm. Then the steel plunger (E) having a snug fit in the cylinder (A) is inserted into (A) and a pressure of 30,000 p.s.i. is applied (note that material for all steel parts is comsteel K 10). If necessary the part (C) can be used to press the sample from (A), and mylar can be used over (B) to prevent sample contamination. Pellets can be numbered with a soft pencil. The amount of sample powder required depends on the nature of the sample but is preferably 1.0–1.5 g. For shorter wave lengths more sample is required to form an infinitely thick specimen. Thus for comparative analytical work it is advantageous to weigh the sample to 1,000 ± 10 mg to achieve equal thickness.

This method of preparing powder samples is essentially that described by Baird and modified by Norrish. (Courtesy K. Norrish, C.S.I.R.O., Australia.)

TABLE XII

SILICATE ANALYSIS, MASS ABSORPTION CORRECTION FACTORS FOR OXIDES OF Fe, Mn, Ti, Ca, K, P, Si, Al, Mg, AND Na IN $Li_2B_4O_7$ + La_2O_3 FLUX (C.S.I.R.O., Div. of Soils, Adelaide, Australia, courtesy K. Norrish)

		Fe_2O_3	MnO	TiO_2	CaO	K_2O	P_2O_5	SiO_2	Al_2O_3	MgO	NaO_2	$Loss$[1]
Fe	1.075	−0.043	−0.065	0.138	0.122	0.127	−0.072	−0.099	−0.106	−0.113	−0.118	−0.167
Mn	1.035	−0.021	0.000	0.203	0.197	0.181	−0.045	−0.060	−0.068	−0.083	−0.083	−0.160
Ti	0.960	0.058	0.030	0.104	0.617	0.625	0.010	−0.015	−0.042	−0.070	−0.096	−0.149
Ca	0.857	0.086	0.082	0.049	0.122	0.737	0.160	0.140	0.103	0.086	0.069	−0.133
K	0.867	0.099	0.081	0.013	−0.012	0.049	0.180	0.159	0.134	0.113	0.080	−0.134
P	0.903	0.054	0.041	−0.036	−0.072	−0.072	−0.100	0.126	0.109	0.091	0.063	−0.138
Si	1.018	0.073	0.085	−0.034	−0.042	−0.064	−0.080	−0.066	0.125	0.099	0.069	−0.158
Al	1.008	0.121	0.113	0.044	0.010	−0.010	−0.028	−0.040	−0.030	0.181	0.167	−0.157
Mg	1.021	0.105	0.096	0.043	0.017	0.000	−0.026	−0.037	−0.055	−0.071	0.154	−0.160

[1] This coefficient indicates a relative weight loss from ignition due to H_2O, CO_2, etc.

$\mathrm{Si_{act}} = \mathrm{Si_{nom}} \times \{1.018 + 10^{-2}[\mathrm{Fe_{nom}} \times (0.073) + \mathrm{Mn_{nom}} \times (0.085)$
$+ \mathrm{Ti_{nom}} \times (-0.034) + \mathrm{Ca_{nom}} \times (-0.042) + \ldots \ldots \mathrm{Na_{nom}} \times$
$(0.069)]\} - B\%,$ \hfill (27)

where, $\mathrm{Si_{act}}$ is actual percentage of SiO_2 sought; $\mathrm{Si_{nom}}$ is the percentage of SiO_2 determined from the calibration curve; $\mathrm{Ti_{nom}}$, $\mathrm{Ca_{nom}}$, etc., are the respective percentages of these elements as oxides as determined from calibration curves, and $B\%$ is background in percent at the particular analytical line determined by using a blank flux pellet.

The nominal values used in the above equations are derived on the basis of a linear relationship between fluorescent intensity, I, and concentration for a (hypothetical) sample with mass absorption coefficient, M.A.C., equal to 1,000.

For such a sample, SiO_2 (nom) $= KI$, where K is a constant converting intensity to %.

Analyzing an actual sample of known composition K can be determined from:

$$SiO_2\% = KI \,(\mathrm{M.A.C.}) - B$$
$$K = (SiO_2 + B)/I(\mathrm{M.A.C.}) \hfill (28)$$

Background, B, is determined by making a pellet with a Si-free sample, whence if I_B is the intensity at the Si K position:

$$B = KI_B \,(\mathrm{M.A.C.}) \hfill (29)$$

K and B must be determined together and if several analyzed samples are used, more reliable value of K will be obtained.

Two actual analyses performed by this method on standard rocks G-1 and W-1 are given in Table XIII. The calculations in this table have been performed by a *computer* but one specific example shall be given here in order to show all individual steps involved, and to demonstrate the tedious nature of mass absorption corrections in X-ray emission analysis.

Let us select, for example, the determination of silica in standard W-1, column 8, lower right corner of Table XIII. The computer has been fed the nominal values of oxides in percent as derived from

TABLE XIII

COMPUTER SHEET DISPLAYING COMPLETE MAJOR ELEMENT ANALYSIS OF STANDARDS G-1 AND W-1 WITH MASS ABSORPTION CORRECTIONS FOR $Li_2B_4O_7 + La_2O_3$ MATRIX (Courtesy K. Norrish, C.S.I.R.O., Div. of Soils, Adelaide, S.A., Australia)

	Fe_2O_3	MnO	TiO_2	CaO	K_2O	P_2O_5	SiO_2	Al_2O_3	MgO	Na_2O[1]	$Loss$	$Total$	$Weight$
Background	0.175	0.165	0.157	0.065	0.054	0.056	0.240	0.125	0.360				
Sample G–1													
Oxide	1.933	0.037	0.253	1.305	5.512	0.080	72.603	14.160	0.460	3.350	0	99.70	0.2800
Element	1.352	0.029	0.152	0.933	4.576	0.035	33.942	7.498	0.278	2.485	0		0.2800
Nominal	2.130	0.204	0.417	1.340	5.500	0.135	73.740	14.530	0.826	3.350	0		
Matrix	0.9898	0.9904	0.9833	1.0224	1.0121	1.0105	0.9878	0.9837	0.9933				
Sample W–1													
Oxide	11.329	0.185	1.121	11.252	0.642	0.140	52.847	15.016	6.682	2.170	0	101.38	0.2800
Element	7.923	0.144	0.672	8.042	0.533	0.061	24.706	7.946	4.031	1.610	0		0.2800
Nominal	11.410	0.347	1.250	11.510	0.702	0.198	52.390	14.930	7.000	2.170	0		
Matrix	1.0082	1.0098	1.0222	0.9832	0.9908	0.9901	1.0133	1.0141	1.0060				

[1] Sodium is determined on a powdered, unfused sample.

calibration curves or intensity ratios. The elements considered are Fe, Mn, Ti, Ca, K, P, Si, Al, Mg, and Na. Oxides of these elements form the bulk of the rock sample being analyzed. The composition of the flux being constant, it is the weight fractions of these elements that affect the total absorption and enhancement effects in the fused sample. Therefore, for silica in rock W-1, we get: SiO_2, W-1 = 52.390 × [1.018 + 10^{-2} × {11.410 × (0.073) + 0.347 × (0.085) + 1.250 × (−0.034) + 11.510 × (0.042) + ... 2.170 × (0.069)}] −0.240 = (52.390 × 1.0133) − 0.240 = 52.847%.

For silica in rock G-1, we get in the same manner: SiO_2, G-1 = (73.740 × 0.9878) − 0.240 = 72.600%.

Background, when expressed as percent of the element determined, is a constant value which can simply be subtracted from the total actual percentage of the element. *The estimation of this background and calibration* are performed for silica, for example, as follows.

(*1*) Determine the background on a pellet made of the plain mixed flux by accumulating, for example, 10,000 counts at 2θ = 109.06°, over a measured period of time. Let us assume we get 16 counts/sec.

(*2*) Determine the total counts per sec at the silicon K_a (109.06°) position, in a fluxed standard, for example, the G-1. Let us assume we get 3,755 counts/sec.

(*3*) Determine the counts per second plus the possible background due to this element's concentration in this matrix by multiplying the total counts by the mass absorption coefficient, M.A.C., which has been previously determined for this rock in the manner shown on pp.273–275. Let us assume this M.A.C. = 0.9883, then 3,755 × 0.9883 = 3,711 counts.

(*4*) Determine the approximate background as follows. Divide the known (actual) SiO_2 percentage of the G-1 sample by M.A.C.: 72.64/C `983 = 73.500% (value not corrected for M.A.C.). This corresponds to: 3,755/73.50 = 51.09 counts/sec/%. Therefore, the *background* of 16 counts/sec in the blank flux corresponds to approximately: 16/51.09 = 0.313% SiO_2.

(*5*) Determine from this the approximate true percentage plus background: (72.64 + 0.313)/0.9883 = 73.81%.

TABLE XIV

CALIBRATION FOR SiO_2 DETERMINATIONS (C.S.I.R.O., Div. of Soils, Adelaide, Australia, Courtesy K. Norrish)

Standard	% SiO_2	M.A.C.*	Fluorescent intensity (counts/sec)	Counts/sec × M.A.C.	$\frac{\% SiO_2 + B}{M.A.C.} = SiO_2 (nom)$**	$\frac{Counts/sec}{SiO_2 (nom)} = counts/sec/\%$
G-1	72.64	0.9883	3,755	3,711	73.8	50.87
W-1	52.64	1.0119	2,669	2,701	52.33	51.01
T-1	62.62	1.0033	3,156	3,166	62.72	50.30
GR	65.90	0.9955	3,338	3,323	66.51	50.18
Mean						50.59

* Calculated from known compositions.
** B = 0.31%.

(6) Determine the counts per second per percent for true percentage plus background only: 3,755/73.81 = 50.87 counts/sec/%.

(7) Repeat this operation for all available *standards*, then take an average as shown in Table XIV. This mean is then used to estimate the *nominal percentages* of silicon in all the calibration curves for the rocks analyzed. For each element, the same procedure with several standards is repeated and the mean counts/sec/% element value used.

(8) After the total counts per second for each element of an *unknown sample* has been determined, divide this value by the mean counts/sec/% value obtained from standards. This gives the *crude value* or *nominal percentage* to be used in the calculation of the total M.A.C. for that unknown rock.

Naturally, when background forms a considerable portion of total counts, as in trace element work for example, this method of background estimation may not be applicable.

CHAPTER 13

Fast-Neutron Activation Analysis

INTRODUCTION

Geochemists usually think of neutron activation analysis as a method specifically suitable to determine trace amounts of elements because the main applications of nuclear activation have in the past been in the concentration ranges of parts per million (p.p.m.) and parts per billion (p.p.b.). Historically, neutron activation dates back to 1936, when Von Hevesy and Levi succeeded in determining one part of dysprosium in a thousand parts of yttrium by bombarding the sample with low-energy neutrons produced in a radium–beryllium source.

This source was a prototype of an *isotopic source*. These are still used today as the cheapest neutron generators and their applicability range has increased due to better supply of different radioactive isotopes. Physically these are simply boxes of shielding material in which the radioactive isotope and the light element, beryllium for example, are combined. A hole through the shielding provides access for the sample to be activated. When a moderator to slow the high-energy neutrons is included, these are called "neutron howitzers". These sources generate slow neutrons. The reactions which produce neutrons in these sources are usually of the (α, n) type. The symbols for alpha and neutron particles, in this sequence in parentheses, denote a nuclear reaction where the alpha particles cause neutrons to be ejected upon collision with nuclei in the sample. In modern isotopic sources plutonium-238 is used, for example, as an alpha emitter instead of radium. Isotopic sources produce relatively low neutron *fluxes* in the 10^3–10^7n/cm^2 sec range.

Neutron activation as an analytical technique has undergone

revolutionary development simultaneously with the fast evolution of the *atomic reactors* during the last twenty years. Presently, the nuclear reactor has to be considered as the *primary source* of neutrons for activation purposes. The energy of most neutrons produced in reactors is relatively low, varying in the 10^{-2}–10^5eV range, and averaging about 220,000 cm/sec in velocity. These neutrons are called "reactor neutrons" and can be subdivided into *thermal* and *epithermal* neutrons—also named collectively *slow neutrons*, and neutrons of intermediate energy of 10^2–10^5eV. Neutrons of energies above 100,000 eV are referred to as *fast neutrons*. Some fast neutrons are also generated in reactors, but they are quickly "thermalized", which means slowed by collisions with atoms in the core and cooling water, or shielding of the reactor. Reactors, as we know, are box-like structures with fissionable cores of some enriched radioactive isotopes or their mixtures. Neutron fluxes, orders of magnitude higher than in isotope sources, are typical for atomic reactors, and they usually cover the neutron flux range from 10^{10}–10^{14} n/cm²sec, depending on the type of reactor. Numerous tubes lead into the cores of these reactors in order to permit the sample transfer for irradiation. Pulsed reactors can give short neutron bursts of 10^{16} n/cm²sec.

The main type of activation reaction in nuclear reactors is denoted by the expression (n, γ) which means that upon collision and interaction between the thermal neutron and the atom, excitation and/or a nuclear transmutation occur, which produces an active isotope emitting gamma radiation. Reactor neutrons constitute the main particles used for trace element work due to the high fluxes available.

A third type of neutron source is the *electrostatic particle accelerator*[1] which uses high voltage potentials between electrodes to accelerate charged particles along a vacuum tube. These particles, usually hydrogen, deuterium, or helium nuclei (protons, deuterons, helions), can produce radioactivity and neutrons when they collide with sufficient energy with light atoms, for example, with the heaviest hydrogen isotope *tritium*, $^3_1$H. Depending on the accelerated

[1] Of Cockcroft-Walton type.

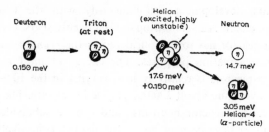

Fig.49. Schematic presentation of the nuclear deuterium–tritium reaction.

particle, we speak of (p, n), (d, n), (α, n) or proton–neutron, deuteron–neutron and alpha–neutron reactions, respectively. The object which receives the impact of these particles is called the *target*. For example, if we have concentrated tritium on a metal plate which is bombarded by deuterons, we name it *tritium target*. In this case, deuterons impinging on tritium knock off one neutron and form an alpha particle by combining in the nucleus:

$$^2_1H + {}^3_1H \rightarrow [^5_2He] \rightarrow {}^4_2He + {}^1_0n + 17.6 \text{ meV}$$

or simply:

$$^3_1H \text{ (d, n) } {}^4_2He + 17.6 \text{ meV}$$

When deuterium is accelerated at 0.150 meV, most of the energy released or about 14.7 meV is carried away by the forward moving neutron, the rest 3.05 meV as kinetic energy, by the alpha particle formed (Fig.49). In an abbreviated form we refer to this reaction as the T(d, n) or simply a (d, n)-type reaction. The very short-lived excited ^5_2He nucleus which exists as an intermediate product as shown in Fig.49 is usually not written into the reaction.

The fast neutrons generated by this reaction have a very narrow energy spectrum peaked near 14.1 meV. Actually, depending on the angle (direction) at which neutrons are scattered from the tritium nucleus in relation to the oncoming radiation, their energies vary from 14.7 meV (forward direction) to 13.5 meV (backward direction), with 14.1 meV energy representing the 90° scatter.

Nuclear transmutation reactions can be spontaneous due to natural radioactivity, as that of uranium or thorium for example, or they can be induced artificially by upsetting the balance within the nucleus with accelerated particles called projectiles, or by high-energy photons. In most cases, a temporary fusion of the projectile with the nucleus first causes a compound nucleus to form, such as the $[^5_2\text{He}]$, for example. The average life of compound nuclei is in the order of 10^{-14}–10^{-12} sec. Compound nuclei decay, therefore, instantaneously by ejection of nucleons, and sometimes positrons or negatrons. The remaining recoil nucleus is usually more stable. Depending on the particle ejected, the elemental nature of the recoil nucleus will change.

Nuclide charts compile the known natural and artificially produced isotopes and give the types of decay of these nuclides. These charts are available from the U. S. Atomic Energy Commission or the Nuclear Research Centre, Karlsruhe, Germany. A key to the location of the products of *radioactive decay* and nuclear transmutation is given in Fig.50. For example, if a proton is ejected, the

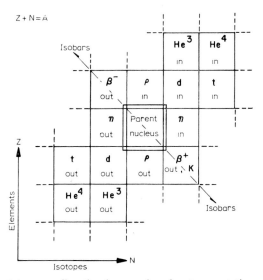

Fig.50. A key to radioactive decay and nuclear transmutation.
A = mass number; N = neutron number; Z = proton or atomic number; He = alpha particle (α); β^- = negatron, electron; β^+ = positron; d = deuteron; n = neutron; p = proton; t = triton; K = electron capture.

position of the parent nucleus will move one atomic number, Z, down. If a projectile proton fuses with the nucleus, the product will be one atomic number higher. When a positron, β^+, is emitted, this results in an electron (negatron) capture, K, and also lowers the atomic number by 1. A negatron, β^-, "out" results in the increase of the atomic number by 1. An alpha particle, α, "out" transmutes the original nucleus to a position 2 atomic numbers lower. In terms of mass, pro ons, p, and neutrons, n, carry one mass unit each and the alpha particle carries 4 mass units. Addition or ejection of these particles changes the mass of the product nucleus by a corresponding number of mass units. Deuterons, d, and tritons, t, carry 2 and 3 mass units, respectively.

In modern *fast-neutron generators*, fast-neutron fluxes range between 10^8 and 10^{11} n/cm²sec. Cockcroft-Walton-type generators accelerate positive particles, Van de Graaff machines accelerate electrons. By choosing proper target materials, neutrons can be produced by both generators. Fast-neutrons interact, especially with light nuclei of such elements as oxygen, fluorine, sodium, magnesium, aluminum, silicon, phosphorus, sulfur, and chlorine. The half-lives ($T_{\frac{1}{2}}$) of the product nuclei of these elements are relatively short, varying from seconds to minutes, whereas the half-lives of heavier isotopes, many of which also interact readily with fast-neutrons, are measured mostly in hours.

The *half-life* of a radioactive isotope denotes the time interval during which half of the radioactive atoms, originally present in a sample, decay. Fast disintegration means short half-life. Half-life has an important connotation for the nuclear chemist because it determines the time necessary to achieve the maximum activation effect with a certain neutron flux. A half-life of ten min, for example, tells us that, if we have some constant neutron flux, half of the nuclei, which under these conditions have a chance to get transformed, will be activated in this time. The total number of product nuclei so formed will depend on the flux and a nuclear factor called *cross-section* (σ), which denotes the probability of capture or interaction between the impinging neutron and the nucleus. Nuclear cross-section can be thought of as a measure of the target area which a nucleus presents to the neutron projectile. One cross-section unit is called a *barn*, it equals, by definition, 10^{-24} cm² per nucleus. Cross-sections for different nuclides vary and they also depend on the energy and type of the activating particle.

The rate of formation, R, of product nuclei per second in a sample is then dependent on the number, N_0, of the stable parent atoms in the sample, on the neutron flux, Φ, in n/cm²sec, and on the cross-section of the reaction in cm², σ. Thus, we can write for the rate of formation of new nuclei:

$$R = N_0 \Phi \sigma \qquad (30)$$

Since the product nuclei are radioactive and decay at a rate determined by their half-life, we can also write for the rate of radioactive decay, D,:

$$D = N^* \lambda \qquad (31)$$

where N^* is the number of radioactive product atoms after a time interval of irradiation, and λ is the *decay constant* of the product nuclide:

$$\lambda = 0.693/T_{\frac{1}{2}}$$

Working with a constant flux we will in time approach an equilibrium when the rate of formation of radioactive nuclei will equal the rate of disintegration, when:

$$R \cong D \qquad (32)$$

At this time, as many new nuclei are created as are decayed. Thus, we have achieved a *saturation point* in time, after which further irradiation with the same flux can not produce more radioactive nuclei and consequently is of no advantage from the activation standpoint. Four half-lives long activation periods, which assure that about 94% of all interactions possible have taken place, are considered as maximum feasible activation time. Saturation has values of $\frac{1}{2}$, $\frac{3}{4}$, $\frac{7}{8}$, and $\frac{15}{16}$ with irradiation times corresponding to 1, 2, 3, and 4 half-lives. Expressed in percent of original nuclide, this is equal to 50, 75, 87.5, and 93.8% yield, respectively.

The rate of radioactive decay, D, can also be expressed by the number, N^*, of disintegrating nuclei per time unit, t:

$$-dN^*/dt = D \qquad (33)$$

or, substituting $N^*\lambda$ for D:

$$-dN^*/dt = N^*\lambda \qquad (34)$$

Integrating we get:

$$N^* = N_0e^{-\lambda t} \qquad (35)$$

where N_0 is the number of nuclei at time zero, $t = 0$, and $e = 2.718...$, base of natural (*Napierian*) logarithms. The exponential decay factor $(e^{-\lambda t})$, or the saturation factor $(1-e^{-\lambda t})$, in the case of activation, result in curves as shown in Fig.51.

The half-life of an activated or natural radioactive isotope is a constant on the basis of which the isotope may be identified. This is important in gamma-ray spectrometry when activities of similar energy, but representing different half-lives, have to be measured and identified. If we plot total counts, N_0, recorded at some starting time zero, $t = 0$, and again after a period corresponding to the half-life of the isotope, and repeat this at intervals of one half-life, we get an identical plot as shown in Fig.51 for the nuclear decay. When we plot the same data on a semilogarithmic scale, a straight decay curve results. The slopes of these curves depend on the half-lives, of the isotopes counted. The shorter the half-life, the

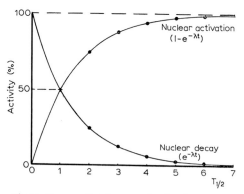

Fig.51. Exponential decay and activation curves.

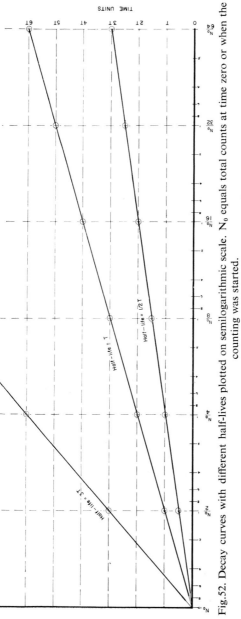

Fig.52. Decay curves with different half-lives plotted on semilogarithmic scale. N_0 equals total counts at time zero or when the counting was started.

steeper the slope, as shown in Fig.52. When activity of interfering isotopes is counted simultaneously by the same counter, and total counts plotted at time intervals, the resulting curves become bent despite the semilogarithmic plot, because the faster decaying nuclide tends to produce a steeper sloped curve than the longer-lived isotope. Since the short-lived component decays faster, its influence on the slope of the decay curve will diminish accordingly, and become insignificant after a time period greater than four half-lives. The longer-lived activity will soon become dominating and the curve straighten out. Curved decay plots on semilogarithmic scale indicate, therefore, contamination and interference of shorter- or longer-lived isotopes than that measured, and from these curves, we can also tell when the effect of the unwanted component has sufficiently disappeared so that counting can start. Obviously, the interference of shorter-lived activity can simply be eliminated by waiting for it to decay. The problem is more difficult in an opposite case when a longer-lived isotope contaminates a shorter-lived which has to be determined. The use of semilogarithmic decay plots (multiscaler decay curves of the multichannel analyzer) is demonstrated in the subchapter on determination of silicon by fast-neutron activation (Fig.76, p.321).

Another important conclusion can be made from eq.30, above. We can see that increasing the flux would increase the rate of formation of radioactive nuclei and thus extend the *detection limits* of the method. In neutron activation, the detection limits are, therefore, directly proportional to the flux. Increasing the flux by one order of magnitude ($\times 10$) usually gives a tenfold lower detection limit.

If we want to calculate the number of radioactive nuclides being formed during one second of irradiation, based on the number, N, of target atoms present, we expand eq.30 by introducing the mass, m, of the activated element in the sample, its atomic weight, W, the *Avogadro* number, A, and the fractional isotopic abundance, f, of the nuclide in question:

$$R = N \Phi \sigma = \frac{mAf\Phi\sigma}{W} \qquad (36)$$

Analytically, short half-life indicates short irradiation time, which is critical when accelerator neutrons are used because the expensive tritium targets when heated by the deuteron beam quickly lose their tritium or decay, p.295. Short half-life also permits *fast analysis*. While thermal neutron activation in atomic reactors is relatively cumbersome and usually requires days, weeks, or even months to achieve the desired activity level, fast-neutron activation of light elements with an accelerator can be accomplished in minutes. Samples activated by fast-neutrons quickly lose their radioactivity and can be handled readily by hand a few minutes after irradiation. Samples activated for long periods in reactors become dangerously radioactive and are difficult to handle or treat chemically. Long irradiations also cause long-lasting radioactivity so that considerable time must elapse before the samples can be further treated or used for other experiments without all the precautions necessary in the handling of strongly radioactive substances. The high cost of atomic reactors prevents their acquisition by relatively small educational institutions or companies, whereas particle accelerators can presently be bought and installed for a comparable sum of money necessary to acquire, for example, a complete X-ray emission or optical grating spectrographic equipment. Utilization and availability of fast-neutron generators has increased manyfold in recent years and their use in *major light element analysis*, as well as in trace element analysis, has become relatively common. The main *advantages* of these methods are speed, the nondestructive nature, applicability to major as well as to trace concentrations, relative independence from matrix due to the high energy of radiation, and selectivity. Fast-neutron activation for major light elements seems to promise the best possible analytical attack when truly nondestructive total analysis without destruction of sample is required. Exceptionally valuable samples which may not even be ground, can be analyzed satisfactorily for many elements by fast-neutron activation. This aspect is especially important in the analysis of rare extraterrestrial material.

Fig.53. Texas Nuclear neutron generator. (Courtesy of the Texas Nuclear Corporation.)

FAST-NEUTRON GENERATORS

Unlike atomic reactors, which even when noncritical, do present radiation hazards and can not be "shut off" in a literal sense because the radioactive core can not be prevented from emitting radiation, the accelerators represent a true means of "push button" activation or "radioactivity on demand", because, aside from the very low energy of the β-radiation emitted by the tritium in the target, radioactivity in the form of fast-neutrons and alpha particles (see reaction p.280) is only produced when the accelerated deuterons are permitted to strike the target.

In the previous chapters, the author has refrained from showing pictures of modern equipment which, after all, usually do not tell the analyst more than that they show mostly impressive streamlined boxes with a multitude of push-buttons. The emphasis has been on the basic principles of operation and thus on simplified dia-

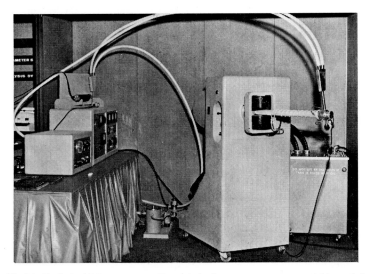

Fig.54. Technical Measurement Corporation's neutron generator (old model) with transfer system tubes, remote control operating console to the left. This picture shows all components of a fast-neutron activation system. (Courtesy of the Technical Measurement Corporation.)

Fig.55. Remote operating, detection, and multichannel equipment in the counting room of the Nevada Mining Analytical Laboratory. Left, Kaman Nuclear remote control console for the Model A-1000 neutron generator, which is shielded behind the wall. Tall black instrument panel is the control of the Philips (Norelco) neutron generator, which operates a sealed tube neutron source. The combination of the Technical Measurement Corporation's Model 404-C multichannel analyzer and integrator is shown on the right. In the middle behind the I.B.M. computer typewriter and, in the left in front, are two 5-inch matched (NaI(Tl)) crystals by Harshaw. Normally these are shielded with lead bricks when used for counting.

grams. While still holding to this principle in this chapter, it was deemed proper to give several photos of different accelerators and associated equipment due to the relative novelty of these machines (Fig.53, 54, 55). These units have to be remotely controlled, well shielded, and the samples to be analyzed have to be transported pneumatically at high speeds through tubes to and from the target of the accelerator. These features are best demonstrated in photographs, floor plans, and logical block diagrams (Fig.56,57).

The *main components* of fast-neutron generators consist of the *ion source* which supplies the desired particles to be accelerated, the

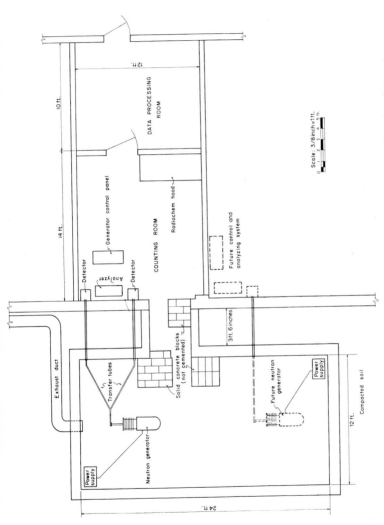

Fig.56. Floor plan for the fast-neutron activation laboratory at the University of Nevada.

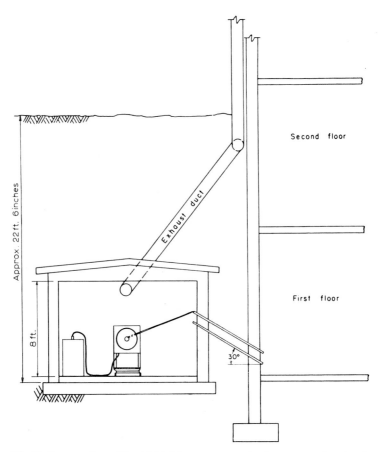

Fig.57. Cross-section of the shielded fast-neutron activation room at the Nevada Mining Analytical Laboratory. Notice the exhaust duct for tritium gas, and the tilted transfer tubes in concrete walls.

Ion source	Accelerating tube		Target assembly
	Focusing of beam	Drift tube	
Production of positive ions		Deflection of ions	Production of neutrons
	Accelerating by different electron potentials	Vacuum pump	

Fig.58. Logical block diagram of an accelerator.

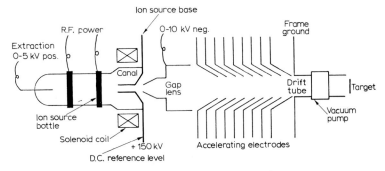

Fig.59. The ion source and the accelerating part of a Texas Nuclear neutron generator. (Courtesy of Texas Nuclear Corporation.)

drift tube with *accelerating electrode assembly* at one end, and the *target* in which the neutrons are produced. Fig.58 presents a block diagram of basic parts of an accelerator.

The *ion source* and the *accelerating* part of the Texas Nuclear Corporation's neutron generator is presented in more detail in Fig. 59. Gas molecules, usually deuterium or hydrogen, are leaked into the *ion source bottle* where they are subjected to *radio frequency* vibrations generated by a power oscillator. Depending on the type of the RF source, about 30–60% of the gas molecules are atomized

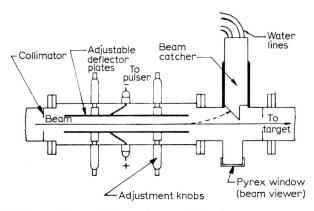

Fig.60. Deflector plate and beam catcher assemblies of a Texas Nuclear neutron generator. (Courtesy of Texas Nuclear Corporation.)

and ionized in this process and, when a potential is applied across this bottle, the deuterium or hydrogen atoms which are stripped of their electrons in this operation are accelerated and escape through a narrow opening in the direction of the drift tube. This process is called *extraction*. The particle beam also contains, therefore, molecular species of the gas. This beam is focused upon leaving the exit canal, and directed through an array of *accelerating electrodes* with increasing potential, for example, in 15-kV steps. Ten such electrodes, each giving the particle a 15-kV kick, will accelerate it to 150-kV energy as the particles will be entering the *drift tube* in which the *collimation* and *deflection* of the accelerated particle beam is accomplished. When the positive particle beam is not deflected toward a negative potential electrode, it strikes the *tritium target* at the end of the drift tube and generates neutrons (cf. Fig.59, 60). The whole generator tube assembly is maintained at a high vacuum between 10^{-5} and 10^{-7} mm Hg.

In principle, all fast-neutron generators are similar. The main differences from the given example above are in ion sources and *beam sorting*. Picker Nuclear neutron generators also use an RF ion source. Other neutron generator manufacturers, for example the Ellison Division[1] of the Technical Measurement Corporation, Kaman Nuclear, division of the Kaman Aircraft Corporation, and the Philips-Norelco Corporation, do use a different ion source named the Penning ion source after its inventor. The Penning Ion Source consists of a cylindrical anode with cathodes at each end (Fig.61). When high voltage is applied to the anode and a strong magnetic field is directed along the axis of the cylindrical container, electrons tend to accelerate toward the anode, in a vacuum of about 1 μ of mercury (1 μ Hg), but are prevented from striking the anode by the magnetic field which bends their paths, forcing them to travel in a spiral way along the axis of the container as shown in Fig.61. The cathodes at both ends of the tube repulse the electrons so that these negative particles are speeding back and forth in this system. Accelerated electrons, as we know, ionize

---

[1] Now extinct, incorporated into Picker Nuclear.

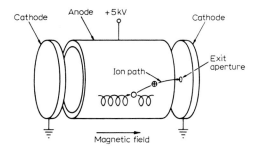

Fig.61. The Penning ion source, of a Kaman neutron generator and the principle of its operation. (Courtesy of Kaman Nuclear.)

gases when they collide with the atoms. Therefore, the bouncing of electrons through the magnetic bottle causes very efficient ionization, and the ionized positive particles tend to move toward the center of the cathodes while being focused by the repulsive forces of the tube-like anode surrounding their path. A pinhole in the center of the cathode directed toward the drift tube permits these positive particles to escape into the accelerating part of the system from which they are speeded toward the target. In this case, the first electrode also functions as an *extractor electrode* by attracting the emerging particles through a high potential of some 50 kV.

Although the *Penning source* provides only about 30% of atomic ions of the resulting beam, it is inherently more stable than the RF source, and due to its rugged and simple construction it also has a long life. When the Penning source is used, more than half of all accelerated particles consist of molecular ions which are much less effective in production of neutrons than the ionized single nuclei. These particles cause unnecessary heating of the target which is objectionable because it speeds the decay of the expensive target without efficient increase of the neutron yield.

To remedy this disadvantage, *beam sorters* are usually supplied with accelerators using the Penning source. These beam sorters work on the same principle as the positive ray apparatus first designed by Thompson in 1912. Presently, Thompson's principle has been developed into highly sophisticated mass analysis. A beam

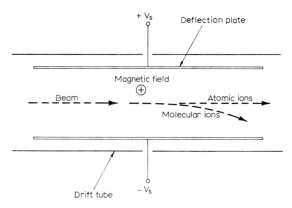

Fig.62.The principle of operation of a beam sorter. (Courtesy of Kaman Nuclear.)

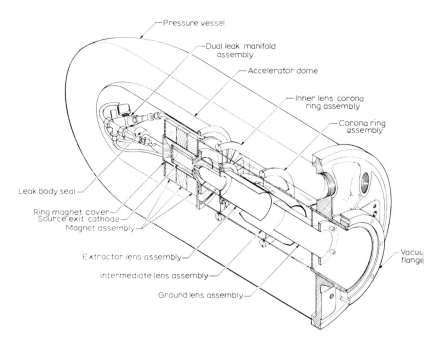

Fig.63. Cross-section of the Kaman Nuclear neutron generator accelerator assembly, and ion source. (Courtesy of Kaman Nuclear.)

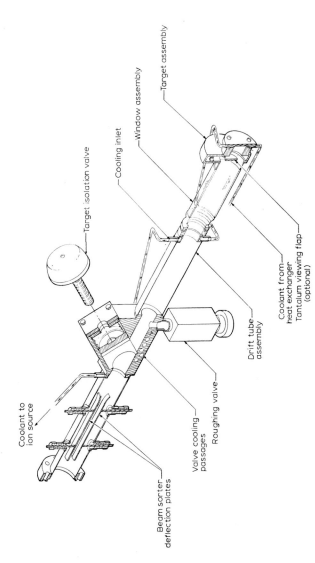

Fig.64. Cross-section of the KamanNuclear neutron generator drift tube beam sorter, and target assembly. (Courtesy of Kaman Nuclear.)

sorter consists of crossed magnetic and electric fields. In such combinations, both energies can be controlled and directed so as to deflect the beam in the desired direction, depending on the energy of the particle, their charge, and mass. The sorter can be adjusted so as to direct the positive ionized single particles toward the target, while deflecting the heavier molecular ions (Fig.62). This is possible because the same potential difference will accelerate light particles more than heavy particles. Cross-section of the Kaman Nuclear accelerator assembly with the Penning ion source is given in Fig.63. Beam sorter and deflection plates of the same make drift-tube assembly are shown in Fig.64.

Both the Penning and the R.F. ion sources work at higher pressures than the rest of the evacuated generator tube and need a steady supply of deuterium gas. Undue increase of deuterium pressure in these ion bottles will affect the low pressure of the system through the narrow pinhole that connects them. Therefore, a well-controlled leak system is necessary to regulate a minute gas flow into the ion source. Usually a *palladium leak*, pictured in Fig.65 is used for this purpose. It is based on the property of platinum metals to become transfluent to hydrogen when heated.

Modern drift-tube neutron generators are supplied with electronic *ion-sputtering vacuum pumping systems*, often called *Vacion pumps*, which is a trade name registered by Varian Associates. Ultek and others also manufacture these pumps. They consist of a titanium cathode and an anode grid structure within a strong

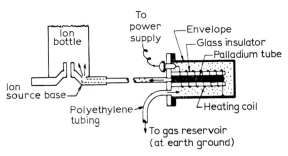

Fig.65. Cross-section of the Texas Nuclear palladium leak assembly. (Courtesy of Texas Nuclear Corporation.)

magnet. When electric current is passed through the anode in a vacuum, the resistance measured is proportional to the vacuum. Noble gases can not be evacuated by these systems.

In order to start pumping, a "roughing" vacuum of at least 10^{-3} mm Hg has to be established with a mechanical pump within the whole system. In this vacuum, further removal of gaseous ions by the sputter pump can start. This is initiated by electrons ejected from the cathode by high energy cosmic radiation. In an electric field caused by a high potential applied between the cathode and the anode structures, the electrons will move in the direction of the anode. The strong magnetic field applied prevents these electrons from reaching the anode and as in the Penning ion source (p.295) they move in spiral paths back and forth between the anodes and are repulsed by the cathodes placed at both ends, at the same time causing ionization of gas molecules. Positive ions thus formed move toward the titanium cathodes and eject titanium atoms from the surface. This process is called *sputtering*. When the sputtered titanium atoms deposit on the anode-grid structure, they are chemically very active and form stable compounds with the gas molecules still remaining in this low-pressure atmosphere.

In this manner, sputtered titanium collects or "scavenges" gas molecules while the high potential and magnetic force are applied to the pump and better and better vacuum results (see Fig.66). The vacuum that can be established with ion sputter pumps is orders of magnitude higher than can be achieved with mechanical or even diffusion pumps. A great advantage of these pumps is the absence of organic substances such as oils, which can cause carbon deposits on accelerator targets when burned. Vacion pumps are noiseless.

The *tritium target assembly* at the end of the drift tube is machined so as to permit easy access and removal of used targets without losing vacuum in the whole system. The target end of the drift tube can be separated from the rest of the evacuated space with an isolation valve as shown in Fig.64. Tritium targets usually consist of a metal layer of titanium or zirconium evaporated first on a thicker metal sheet, which is usually copper, and then tritiated. The

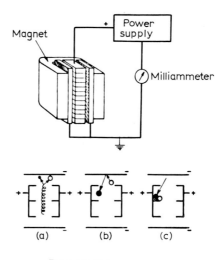

- • Free electron
- o Gas molecule
- ● Titanium atom

Fig.66. The principle of operation of a Vacion pump. (Courtesy of Texas Nuclear Corporation.)

tritiating of the thin metal layer is performed by heating and cooling the target sheets in a tritium atmosphere at low pressure.

Changing of targets, which often carry about 10 curies (Ci = $= 3.7 \cdot 10^{10}$ disintegrations/sec) of radioactivity as low-energy beta (β) emission associated with the decay of tritium atoms, is a potentially *dangerous* procedure to be performed wearing rubber gloves and goggles. Tritium gas is poisonous and its low-energy beta-radiation is harmful to the human body and lungs when inhaled. Tritium, being hydrogen, penetrates directly through the skin and is easily adsorbed on metal surfaces. These dangers are compounded by the fact that the very low-energy (0.019 meV) beta emission of tritium is difficult to detect and special *tritium detectors* are necessary to guard and monitor against possible tritium contamination. The half-life of tritium is 12.4 years, and its *biological life*, by which we denote the time period the human body takes to rid itself of this radioactive isotope, is about two weeks. Because of this, urine tests for tritium are mandatory for workers changing tritium targets in accelerators.

The above described *drift-tube type neutron generators* are capable of maximum fast-neutron yields of about 10^{11} n/sec when operated with new tritium targets at full capacity of about 200 kV, and a 1-mA beam current. However, the operation under these maximum conditions is very expensive because targets, which may cost two hundred dollars apiece, would last only about 20–30 min. Normally, in regular analytical routine, one uses fluxes of about 10^7–10^8 n/cm²sec. When fluxes and yields are discussed, it is important to distinguish exactly what is meant by these terms.

A total *yield* or *output* corresponds to the number of all the neutrons generated from a point-source in 1 sec. Conventionally, a tritium target is also considered a point-source when total output data for an accelerator are given.

The yield is expressed in neutrons per second (n/sec).

A *flux* from an isotropic source corresponds to the number of neutrons per unit area normal to direction of neutrons at a distance, r, from the source. The flux is expressed in neutrons per square centimeter per second (n/cm²sec).

To understand the numerical relationship of the flux to the yield, we must recall the $T(d, n)$ reaction on p.280. Modified, we can represent the same reaction in a geometric system as in Fig.67.

In Fig.67 the neutron flux is dependent on the steradian (circled area) which is defined by the angle $\ominus$ and the radius, r, or the distance from the point source, which is idealized as a single target nucleus. In nuclear reactions where the angular scattering of product nucleons approaches isotropic distribution, the number of such particles is independent of the angle θ. Therefore, we can write for the total number, N, of neutrons per steradian per second corresponding to an angle $\ominus$:

$$N = Y/4\pi \text{ n/steradian/sec} \qquad (37)$$

where Y is the total yield of neutrons per second, and in order to calculate the flux, Φ, at a certain distance, r, from the center of a target, we write:

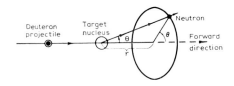

Fig.67. Isotropic scattering of neutrons from a target nucleus. Neutron flux per steradian depends on the angle θ if the angular distribution is isotropic and, thus, independent of the angle θ. This is essentially true for the $T(d,n)$ reaction at 150 kV acceleration energy.

$$\Phi = Y/4\pi\,r^2 \tag{38}$$

Thus, for an isotropic target source emitting $8 \cdot 10^{10}$ n/sec, we can calculate the flux, for example, at a distance of 2 cm from the center:

$$\Phi = \frac{8 \cdot 10^{10}}{4 \times 3.14 \times 2^2} = 1.6 \cdot 10^9 \text{ n/cm}^2\text{sec}$$

Examination of eq.38 shows that in order to utilize the maximum flux, one has to place the sample as close to the target as possible. Strictly, the T (d, n) He reaction is not isotropic, because the flux is about 15% less in the backward direction.

Evaluating the main *disadvantages* of drift tube accelerators, we find that the troublesome target change with the precautions necessary, and the re-establishment of high vacuum, as well as the health hazards involved, constitute major obstacles in establishing and running these machines routinely in laboratories ill-equipped for handling dangerously high levels of radioactivity in terms of trained personnel and tritium monitoring. In order to alleviate these difficulties and provide the analyst with a truly push-button type neutron source of maximum safety and simplicity of operation, several manufacturers build *sealed neutron sources* or tubes which seem to gain acceptance by the chemist and may replace the drift tube generators in routine analytical work.

Sealed neutron tubes basically consist of the same components as the drift-tube assemblies. There is an ion source, a short focusing and accelerating section, and a target built into a sealed evacuated chamber. The main difficulty in building sealed neutron tubes economically results from the inherently short life of tritiated targets. It is cheaper to replace a demountable target in a drift-tube machine than to throw the sealed assembly away every time a target is depleted. Therefore, successful attempts have been made to build self-replenishing sealed tube assemblies which work on the principle that when tritium is accelerated with deuterium it partially adheres to the target and compensates for the losses due to the heat generated by the impinging beam. One of the very first successful *selfreplenishing neutron tubes* was built by the Philips Corporation. In this tube, the replenishing by tritium is performed by heating a tritium-saturated wire which releases tritium into the low pressure system of the tube. Such tubes can deliver uniform neutron yields up to 10^{10} n/sec for periods over 200 h.

More recently, other companies have manufacturered sealed

Fig.68. The Kaman Nuclear sealed neutron tube. Maximum output 10^{10} neutrons. (Courtesy of Kaman Nuclear.)

neutron generator tubes with yields up to 10^{11} n/sec, but with a somewhat shorter life (~ 100 h). Such a tube built by Kaman Nuclear is shown in Fig.68. It may be forecast that the simplicity, relative safety, and the small space required for sealed neutron generator tubes will make their use for analytical purposes more popular and widespread than that of the drift-tube type machines, which, however, will doubtlessly retain their place as more versatile units in the hands of physicists and when ultimate fluxes are desired in trace element work.

Laboratories shielded for fast-neutron activation analysis

Typically, neutron activation analysis is performed by remote control to protect the personnel from the hazards of nuclear radiation. Sources of this radiation are shielded by different materials which are selected to slow down (moderate) the particles and stop as much of the radiation produced by the radioactive decay or activation as possible. While alpha particles can be stopped by relatively thin sheets made of almost any material, and even high-energy betas can be stopped by thin sheets of heavy metals such as iron or lead for example, gamma radiation, especially in the meV-energy range, can penetrate thick sheets of metal. Neutrons present exceptionally difficult shielding problems because they move freely even through heavy metal shielding and in all directions while being bounced by collisions with nuclei and slowed down. As we have seen above (p.297), reactor neutrons possess a wide energy spectrum and propagate, therefore, at different velocities, finally moving at speeds of 2,200 m/sec.

All neutrons are ultimately captured by atoms, but their paths are unpredictable and being neutral particles they can not be detected directly. Fast-neutrons possessing high energy and speeds of $5 \cdot 10^9$ cm/sec, can travel hundreds of meters from the source before being captured. Neutrons in general are best slowed down by hydrogen and light elements. Hydrogen is most effective in neutron moderation. An average energy loss per elastic collision is one half. It is one quarter for carbon and less for heavier elements.

For fast-neutrons, water, paraffin, or liquid hydrocarbons, and all hydrogen-rich substances, are therefore the best shielding material. Concrete, which usually contains about 20% water, and plain clay-rich soil, are also used for shielding materials. For building vaults for fast-neutron generators, concrete is most suitable due to its general availability and low cost. From eq.38 we can also see that distance from the source is one of the main factors which must be taken into account in shielding. For example, a fast-neutron flux at 100 m distance from a source emitting a total of 10^{10} n/sec, will be about: $10^{10}/12.57 \cdot 10^8 = 8$ n/cm²sec assuming that shielding effects are absent.

The *maximum permissible whole body exposure* for personnel for a 40 h working week is 10 n/cm²sec. This corresponds to 100 milli-röntgen equivalents per man (100 m.r.e.m.) in a 40 h week period. Thus, in our case, if we could work at a distance of 100 m from the source, we would not need shielding if we spent only 40 h/week with the generator operating. It has been experimentally determined that about 40 cm of concrete decrease the fast-neutron flux by approximately a factor of 10, so that if we provided concrete shielding 320 cm thick, the same effect would be achieved. Since this implies a distance from the source of 320 cm, this would further reduce the flux by four orders of magnitude, $320^2 \simeq 10^4$, which means that if we know that we will be working at this distance from the target, we can reduce the thickness of the concrete wall by 160 cm to a thickness of 1.6 m or about 5 ft. It must be remembered, however, that this concrete thickness is only sufficient as long as the worker remains at a distance of at least 320 cm from the target at all times. Exact calculations are difficult due to the nature of neutrons.

It is recommended that when shielding is planned, the *maximum possible fluxes* and *continuous exposure of the personnel* are considered as normal conditions in the laboratory, and that whenever economically feasible, an additional safety factor in terms of added concrete or shielding is built in. A scientist will often spend long hours exceeding the total forty hours period per week in a laboratory when developing analytical methods. In addition to the initial equipment, other radioactive sources may be installed later and tested at maximum neutron output conditions simultaneously with the originally installed accelerators which may result in higher neutron levels than permissible, or different neutron energies.

When shielding is considered, one must check in all directions for sufficient concrete thickness and distance from the source. The area above and below the generator is just as critical as the thickness of the walls. If no special vault can be provided, a corner in a basement or a hole underground through the concrete walls are the simplest solution. Whenever the generator's drift tube is projected into a *cavity* in the wall, one must also provide shielding *behind* the

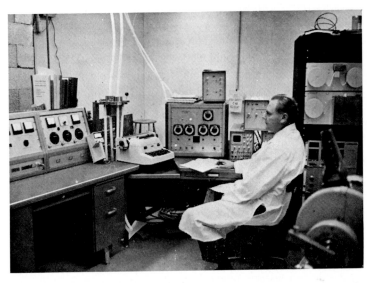

Fig.69. The complete detection, sample transfer, and data processing system for fast-neutron activation at the Nevada Mining Analytical Laboratory. Left: The Kaman remote control console for the Kaman Model A-1000 neutron generator. Behind it, piled bricks shielding the entrance into the vault. White plastic tubes lead to a dual transfer system, built by Technical Measurement Corporation, with a sample switching table to permit consecutive irradiation and counting in alternate tubes. I.B.M. computer typewriter, T.M.C. console built for this laboratory with clocks for timing the transfer, irradiation, and counting, in the middle. T.M.C. 404-C multichannel analyzer and integrator behind the person. Tally computer tape perforator and reader for fast accumulation of spectral data, on the right.

Fig.70. Kaman Nuclear sealed tube neutron generator, shield of paraffin, and concrete pit in foreground. Installation of Kaman Nuclear facilitie., Colorado Springs, Colorado. (Courtesy of Kaman Nuclear.)

generator, remembering that the neutron output is isotropic. In any case, monitoring for neutron radiation all around the area enclosing the generator is mandatory irrespective of shielding thickness or material used. Bricks, often used for these purposes, may accidentally be stacked in such a way that holes or gaps exist.

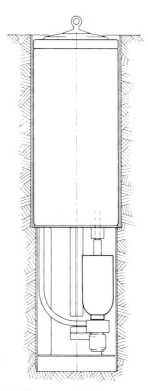

Fig.71. Cross-section of the Kaman sealed tube neutron generator installed in a pit with a removable block of paraffin as shielding plug on top. Compare with Fig.70. (Courtesy of Kaman Nuclear.)

Horizontal tubes for sample transfer systems can also serve as channels for neutrons if the target happens to be at their height. All possible openings and electric conduit tubes must also be checked. Since conditions in any laboratory usually change with new experiments and personnel, neutron monitoring checks have to be performed periodically. The building of new structures in the vicinity of activation vaults has to be restricted and closely controlled.

The physical size of generators determines the type of shielded vault best suited for the purpose. *Drift-tube generators* require easy access, especially to the target assembly. They are usually built on wheels to permit their positioning in atomic reactors and to enable the user to achieve extra shielding by pushing the drift tube end into an orifice in concrete walls. An ideal case where a separate vault could be built underground is shown in Fig.56 and 57. The fast-neutron activation laboratory at the Mackay School of Mines, Nevada Mining Analytical Laboratory, University of Nevada, was planned and designed by the author to permit total nondestructive fast-neutron activation with two neutron generators (one a drift-tube type and the other a sealed tube type) combined with light element analysis by X-ray emission, with a computerized output. Fig.55 and Fig.69 demonstrate the electronic equipment to process the data and the remote controls with a dual transfer system described in detail below.

Sealed tube neutron generators providing hundreds of hours of continuous operation may be lowered into a *concrete well* or *pit* on the ground floor of the building, as shown in Fig.70 and 71.

This shielding method may prove to be the most economical in small laboratories built before fast-neutron activation was contemplated. Also, in this case, a sufficient concrete, or a paraffin block which is light to lift, must be provided.

DETERMINATION OF OXYGEN BY FAST-NEUTRON ACTIVATION

Oxygen is the major element in the earth's crust and ranks third in cosmic abundance after hydrogen and helium. The stability of the oxygen-16 isotope is explained by the structure of this nuclide. It consists of eight protons and eight neutrons, thus incorporating in its nucleus two "magic" or special numbers which means that both the proton and the neutron shells are filled, resulting in an exceptionally stable nucleus. Three stable oxygen isotopes occur in nature. The oxygen-16 isotope is much more abundant than the oxygen-17 and the oxygen-18 isotopes. According to Nier, the relative abundances of natural oxygen isotopes are: $^{16}O = 99.759\%$; $^{17}O = 0.0374\%$; $^{18}O = 0.2039\%$.

In igneous rocks, oxygen averages 46.6% by weight. The hydrosphere contains 86%, the atmosphere 23%, and meteorites 32.3% oxygen by weight, according to Goldschmidt. The average weight percentages of iron, oxygen, and silicon on the planet earth are 38.8%, 27.2%, and 13.8%, consecutively.

Chemically, oxygen is the main negative ion (anion) in most materials. Oxygen, element atomic number 8, stands between nitrogen and fluorine in the sixth (VI) main group of the second period of the periodic system of elements. Most oxidation reduction processes are directly, or indirectly, based on reactions with oxygen. All silicates, carbonates, sulfates, phosphates, and oxides are dominantly oxygen compounds. The weight of oxygen in these compounds varies in the vicinity of 50%, yet until recently a chemist was rarely concerned with the problem of *oxygen analysis*. This is explained by the fact that in most analyses of oxidized compounds the final results are conventionally reported in oxide percentages. In all cases where other negative ions may substitute for oxygen and in cases where multiple valences occur as well as in cases of incomplete oxidation of a fraction of cations or non-metals, ambiguities, and stoichiometric problems have arisen. In all such cases, the reporting of cations in oxide form can lead to a fictitious total percentage of more than one hundred.

The main difficulties in conducting an oxygen analysis result from the oxidative effect of the atmosphere and the reducing effect of metallic or organic substances. Flames and furnace atmospheres are reducing in general unless special precautions are taken. Therefore, classical oxygen analysis has to be performed in a vacuum, and, due to the strongly negative character of the oxygen ion, only fluorine compounds can be used to liberate oxygen from most of the inorganic substances. Strongly electropositive metals can also be reacted with oxygen in order to separate it from its original environment. All these methods of oxygen separation and analysis are cumbersome and time consuming so that the analyst has generally accepted the conventional approach and reported his analytical results in *percent oxide* and introduced corrections only when the known composition made it necessary, for example for fluorine, F, sulfur, S, and chlorine, Cl. In order to complete and calculate a total analysis of a rock or mineral properly, we actually need an independent analysis of oxygen.

Fast-neutron activation analysis of oxygen is the most suitable and interference-free analytical method known for this element in all kinds of mixtures. Due to the great stability of the oxygen atom, particle energies needed to penetrate and disturb its nucleus are exceptionally high. Consequently, this results in a highly unstable activation product with very high energy gamma radiation. Only a few radioactive isotopes emit similar gamma-energies, which makes fast-neutron activation for oxygen an exceptionally interference-free method. When an oxygen-16 isotope is activated by 14-meV neutrons the nuclear reaction that takes place is denoted by: $^{16}O(n, p)$ ^{16}N. When an oxygen-16 nucleus is hit by a fast-neutron of about 14 meV energy, a proton is liberated and a highly unstable nitrogen-16 isotope is formed. This nitrogen decays by emitting high-energy betas to an excited ^{16}O, which emits gamma radiation of 6.13- and 7.12-meV energy. Note that, due to the high energy of the "nitrogen" gammas, counting with small crystals may produce higher escape peaks than the primary radiation.

Only the reaction: ^{19}F $(n, \alpha)^{16}N$ produces the same isotope. Nuclear reactions producing the same nuclide are called *primary interfering reactions*. Another interfering reaction to be considered, but only in rocks or samples with major boron, is: $^{11}B(n, p)^{11}Be$ because the beryllium-11 isotope is a high-energy beta-emitter with about double the half life of the nitrogen-16 isotope. When ^{11}Be decays back to ^{11}B, this isotope is first in an excited state and emits gamma radiation at 7.99, 6.81, and 1.98 meV. The low cross-section for this reaction and the low branching ratio of the 7.99 and 6.81 gamma rays only make this interference significant when boron exceeds oxygen by a factor of about one hundred, in the sample.

The half-life of ^{16}N is 7.13 sec which means that in order to be counted the activated sample has to be brought quickly to the detector after activation.

Despite the fact that we work with an oxygen isotope instead of the oxygen as an element, knowing that the relative abundances of oxygen isotopes vary little and considering that the oxygen-16 isotope represents 99.76% of the total isotopic composition, we can assume that the determination of the gamma intensities of this dominating oxygen isotope will give us a satisfactory relative figure on the total concentration of the oxygen element in the mixture.

Fast sample transfer to and from the target is a critical condition for precise oxygen determination. From the short half-life, we can see that in 7.13 sec one half of all activated oxygen-16 nuclei will be decayed. Thus, pneumatic transfer systems capable of carrying the samples from the generator in less than 1 sec have been built by several institutes and companies interested in fast-neutron activation. Most of these systems constitute a *single tube transfer channel* which permits *consecutive* irradiation and counting of samples only. However, working with neutron generators, one soon discovers that the deuteron beam wanders over the target and that its intensity and the intensity of the resulting neutron radiation are relatively *unstable*.

If the total amount of neutrons emitted were known for the irradiation time, the analyst could calculate the number of neutrons passing through the sample, knowing the nuclear cross-section, the flux, the distance, the sample size, etc. Therefore, in *absolute* nuclear assay techniques the analyst needs to monitor the total number of neutrons emitted during an activation.

In *comparative* nuclear assay techniques, the activated sample is compared to a sample of similar known composition and the concentration of the analyzed isotope is estimated by comparing the relative intensities. The latter method is applied here because it is analytically more straightforward and more suitable for analyses of mixtures. Yet we find that even using comparative methods we do not eliminate the effect of varying total neutron fluxes received by the sample and standard unless both have been subjected to the same fluctuations of the flux for the same time. This can be achieved by *simultaneous irradiation* of sample and standard in a transfer system that permits parallel conveying of two containers to and from the irradiation facility. Such a system can be called *dual transfer system*. When both samples are subjected to simultaneous fluctuating neutron output for the same length of time, the need for expensive monitoring in order to correct for different total neutron fluxes in consecutively irradiated samples does not exist. It is true that ingenious flux monitoring and correction methods have been designed recently for single tube transfer systems but an analyst seeking simplicity of operation and reliability is well advised to start with a dual transfer system.

Precision is also affected in fast-neutron activation by *irregular tritium targets* which cause "spotty" neutron distribution in front of the target assembly so that one side or one part of the sample gets more neutrons than the other. This because the distribution of tritium in a target is similar to the distribution of photosensitivity of a photomultiplier tube shown in Fig.72. This is corrected by *rotating* the sample capsule or "rabbit", as these fast-moving sample containers are often called. To be effective, this rotation is best performed around two axes as shown in Fig.73. Another effective method is to change samples consecutively from one transfer tube

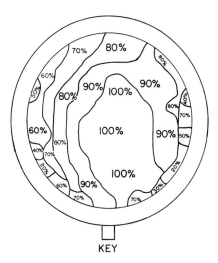

Fig.72. Irregular sensitivity distribution of a RCA 2-inch diameter photomultiplier tube, type 6342A. When samples are counted at different locations and the intensities compared, a bias may result. This bias can be corrected by shifting the samples consecutively from one counting tube to the other as is done at the Nevada Mining Analytical Laboratory. (Courtesy W. C. Kaiser, Argonne National Laboratory.)

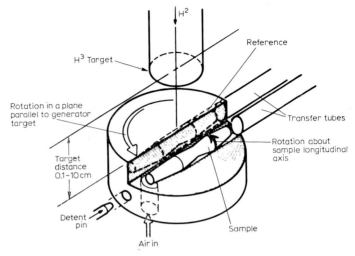

Fig.73. Double axis rotator which equalizes the effect of "spotty" targets. (Courtesy of Kaman Nuclear.)

into another, as is done in the Nevada Mining Analytical Laboratory.

A less serious effect on accuracy can be caused by irregular sensitivity of the photomultiplier tube (Fig.72) if irradiated samples are permitted to be counted in different positions in relation to the crystal detector.

Precision is also affected by *drifts* common to all electronic equipment and, in the case of accelerator-produced neutrons, precision is greatly affected by the decay of targets which results in lower counting rates with each consecutive irradiation. In single transfer systems, these interferences are very serious and have to be corrected in accurate work. In dual transfer systems, an *internal automatic correction* occurs when a comparative analysis is performed because intensity ratios of two samples rather than absolute intensities are recorded.

A *dual transfer system* demands two parallel detector and counting systems. When two separate electronic systems are used, electronic noise and background levels may be different and a characteristic of parallel electronic systems in general is their tendency to display *cross-drift*, which typifies changes in output which are not parallel, so that while one amplifier's noise grows or electronic window shifts slowly, the other detector output may grow slower or actually fall with time of operation.

In a comparative system, this reflects on the accuracy of results which will drift further apart or become more comparable in time, while the analysis is being conducted. For an analyst, this would mean that relative data retrieved one day may not be of the same confidence level the next day. The multitude of electronic equipment connected in a series for counting and registering the pulses in fast-neutron activation is such that it is highly improbable that all the electronic components could be regulated or controlled so as not to drift apart in time. In order to minimize this effect the Nevada Mining Analytical Laboratory is using a system designed to switch the samples after each transfer so that they are irradiated and counted consecutively in opposite locations. Fig.74 shows the logic of this system in a block diagram. The internal correction in this system results from consecutive accumulation of counts originating in different activation-to-detection combinations. When two of the counts obtained from two different "detector"–"activation spot" combinations are summed, an internal correction for the possible differences in background or drift results. If counting rates in both combinations are similar, this correction will be

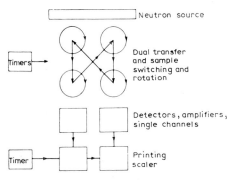

Fig.74. Logical diagram of the dual self-correcting transfer system at the Nevada Mining Analytical Laboratory.

sufficient. If counting rates are different, the combination of further irradiation and switching progressively improves the reliability of results. This system has produced remarkable precision and accuracy. The comparative freedom of elemental and matrix interferences in this case also results in improving accuracy with the increasing number of averaged determinations.

A dual transfer system with a switching mechanism permitting consecutive irradiation and counting in different channels, also corrects for the irregularity in targets, and the sensitivity differences of the photomultiplier tube (Fig.72).

Procedure

Sample powders are *packed* into plastic pre-weighed screw-top containers or "rabbits" (Fig.43) of known low oxygen content. The packing has to be done tightly in order to avoid slushing effects of the powder. This is achieved by packing small quantities consecutively and compacting with a small flat-top pestle. The plastic screw-top is then tightly closed, the rabbit cleaned from outside and weighed. In accurate work, the effect of *adsorbed water* ($-H_2O$) has to be considered since water contains nearly 89% oxygen. In that case, sample powder is kept for 2 h at 105°C, then cooled in a dessicator, and packed immediately. Standards are handled similarly. In this laboratory, about 2 g of rock powder are packed in a "rabbit" of 1-cm³ capacity, and it must be kept in mind that in a large sample the effect of atmospheric or container oxygen contamination may be ignored. With small samples these factors become significant and packing has to be done in oxygen-free atmosphere.

The maximum irradiation time for oxygen is 30 sec, because the half-life of the product ^{16}N is 7.3 sec. The saturation time for four half-lives, $4T_{\frac{1}{2}}$, is thus 29.2 sec. The standard and the unknown samples are propelled pneumatically to the irradiation point in front of the target and after their arrival has been ascertained, irradiated for 30 sec, and returned to be counted in front of two separately shielded scintillation crystals with phototubes. Counting time is also maintained at 30 sec because of similar considerations as above. This procedure is repeated at least twice. For accurate results, eight irradiations are necessary, remembering that every consecutive count on a sample will be accumulated in the alternate channel. Two counts from different counters can be considered as one complete corrected determination when added.

Making sure that samples are at rest in front of the respective target and crystal is important because fast-moving "rabbits" have been observed to bounce a few times after the first impact. This is best accomplished by allowing enough time after transfer before activating or starting the count. All these consecutive steps are

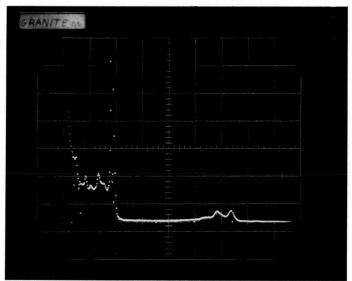

Fig.75. Gamma-ray spectrum of a granite. Two peaks in the lower right quadrangle are ^{16}N gamma peaks at 7.13 and 6.13 meV. The sharp high peak in the center of the first half of the CRT screen is ^{28}Al gamma peak at 1.78 meV which represents silicon in the sample. On the left, the relatively high annihilation peak at 0.511 meV, and, between it and the ^{28}Al peak, other superimposed peaks representing ^{27}Mg at 0.84 and 1.04 meV, from reaction ^{27}Al(n,p)^{27}Mg, and ^{56}Mn overlapping at 0.845 meV, from the reaction ^{56}Fe(n,p)^{56}Mn. For better resolution in this region, semiconductor detectors and mega channel analyzers can be used. When small crystals are used, the ^{16}N escape peaks dominate often, which must be considered when the spectrum is calibrated. (Courtesy of H. A. Vincent.)

timed by clocks and controlled electronically as shown in Fig.69. The dual amplifiers and single channel analyzers are so set as to accept all gamma energies between 4 and 8 meV which represent oxygen. The ^{16}N spectrum is energy wise so far apart from other gamma lines, even in a complex silicate rock gamma-ray spectrum (see Fig.75) that in this particular case the use of a wide window *single channel analyzer* is permissible in nondestructive analysis. More accurate work can be done, and the characteristics of the nitrogen spectrum and the other parts of the composite gamma-ray spectrum examined, if one uses multichannel analyzers briefly described on pages 220–227, 290 and 306. The counts under the oxygen peaks are registered from single channels or integration is performed electronically if a multichannel analyzer is used, and their ratio taken. This ratio closely represents the oxygen ratio in a self-correcting system, such as described here. For even more accurate work, this ratio can be plotted and unknown samples bracketed between known standards of similar composition. If the percentage of oxygen in the standards is known, multiplication by the count-intensity ratio will give the percentage of oxygen in the sample.

In *single tube transfer systems*, or when samples are not switched, the intensity ratio can not be used as such to calculate the oxygen percentage in the unknown because of drifts and background differences in these parallel systems due to electronic instability. In such cases, consecutive irradiation in the sequence standard—sample—standard is recommended and plotting of a working curve similar to that in X-ray emission analysis is necessary. Table XV gives a series of results achieved with the self-correcting oxygen analysis system at the Nevada Mining Analytical Laboratory. These results are based on relative counting intensities and were obtained directly without calibration curves.

In all work with very short-lived isotopes, the problem of *dead time* (p. 220) arises, despite the internal correction for this phenomenon built-in in most modern multichannel analyzers. This is due to the fact that by the time the analyzer starts correcting for the missed time, the activity has practically decayed so that there

TABLE XV

PRECISION AND ACCURACY OF OXYGEN ANALYSIS BY FAST-NEUTRON ACTIVATION, SiO_2

Sample	Theor. oxygen (mg)	Oxygen det.d. (mg)	Ratio of oxygen in samples	Ratio of oxygen det.d.	% of oxygen theor.	% of oxygen det.d.	Dev. from theor. perc.	No. of det.ns	Relat. dev. (%)	Min. counts recorded per det.n.
SiO₂, quartz, ignited:										
Nevada Quartz, no.1	779.6	781.2	0.9547	0.9567	53.25	53.36	+0.11	32	1.98	57,000– 90,000
		780.7		0.9560	53.25	53.33	+0.08	32	1.93	60,000– 90,000
		784.3		0.9604	53.25	53.58	+0.33	8	1.59	73,000– 84,000
Nevada Quartz, no.3	816.6	814.9	1.0474	1.0453						
		815.5		1.0460						
		811.7		1.0412						
Quartz, 99.89%, no.5	1129.3	129.8	0.9985	0.9985	53.25	53.25	±0.00	20	1.88	76,000–135,000
		(1,127.7)		0.9966	53.25	53.18	−0.07	32	(1.86)	76,000–135,000
		1,127.7		0.9966	53.25	53.18	−0.07	12	2.09	119,000–174,000
		1,127.2		0.9962	53.25	53.16	−0.09	4	2.78	250,000–265,000
Quartz, 99.89%, no.10	1130.9	1,130.9	1.0015	1.0015						
		1,133.1		1.0034						
		1,133.1		1.0034						
		1,133.6		1.0038						

are no counts incoming which are due to the desired nuclide. This can be alleviated by selecting similar standards or activating for shorter periods and running through more cycles. The dead-time effect increases exponentially with the increasing counting rate.

DETERMINATION OF SILICON BY FAST-NEUTRON ACTIVATION

When we examine the composite gamma-ray spectrum of a silicate rock over the 0.1- and 8.0-meV energy region, we can usually detect two major features. These are the nitrogen-16 or "oxygen" peaks at 7.13 and 6.13 meV with their respective escape peaks at energies lower by 0.511 meV each, and a sharp peak at 1.78 meV (Fig.75). This peak corresponds to gamma radiation originating from aluminum-28 produced in the reaction $^{28}Si(n,p)^{28}Al$, induced by the fast-neutrons. In this reaction, a fast-neutron "fuses" with the silicon nucleus, knocking out a proton. Aluminum-28 decays back to silicon-28 by emitting a negatron (β^-) and gamma radiation at 1.78 meV, with a half-life of 2.3 min.

As in the case of oxygen, silicon has only three natural stable isotopes of which the one with mass 28 is dominating with 92.2 wt. %. The isotopic abundance of silicon varies in rocks within such narrow limits that no correction for this possible effect is necessary.

On the high-energy or the right side of the aluminum or "silicon" peak is the fast-decaying "oxygen" spectrum, on the left side the slower decaying "iron" peak at 0.845 meV. This is due to decaying manganese-56 from the reaction $^{56}Fe(n,p)^{56}Mn$. Under these conditions, if we wait 4–7 half-lives of nitrogen, or 30–60 sec, the short-lived component of the spectrum will be practically eliminated. If we now count the aluminum-28 activity, there will be no interference from "oxygen", but there may be interference from other longer-lived active isotopes formed during the irradiation. Plotting counts collected at intervals over the aluminum peak versus half-life will then reveal other decaying components. Shorter-lived isotopes will produce a steep curving slope at the beginning end of the semilogarithmic plot, longer half-lives will tend to flatten the decay curve tail. In analytical work, one selects a time period during which the decay curve is linear on the semilogarithmic plot of the decay. In comparative activation analysis of a sample, the total counts taken in the time interval corresponding to this linear region are related to total counts of the standard recorded over the same relative time period. This ratio is then used to calculate the composition of the unknown. Half-lives are thus used to select a most interference-free time interval representing the activity measured and at the same time as an additional means of identification (see Fig.76).

Procedure

Prepare the sample and standard and activate as described under Oxygen; transfer, and count under multiscaler mode of the multi-channel analyzer, collecting both activities in separate memory halves. Determine the effect of interfering half-lives and from it the waiting period necessary in order to let the "oxygen" and other short-lived activity decay. Plot decay curves of sample and standard

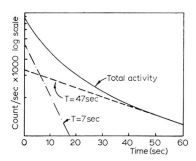

Fig.76. The effect of decay of short half-life radionuclide on a decay curve of a longer half-life when measured simultaneously. In the case shown, the composite curve approaches linearity at about 50 sec after the start of the counting. The short-lived isotope has a half-life of 7 sec, corresponding closely to the half-life of ^{16}N. We can thus see that if we want the "oxygen" activity to decay at least 99%, we have to wait about 50 sec, preferably 60, when the curve becomes linear. Counts taken after that waiting period would represent the nuclide with the half-life of 47 sec. (Reproduced after a similar drawing by Texas Nuclear.)

on a X-Y recorder and determine the linear decay period common to both samples (Fig.76). This can also be done visually from the Cathode Ray Tube (C.R.T.) display or an oscilloscope. Activate again, wait the time period determined, which in this case will be between 60 and 90 sec, and accumulate simultaneously in two different memory halves of the multichannel analyzer. Integrate all counts under respective peaks over the same number of selected channels, and compare the relative intensities directly, or use calibration curves, if necessary. Repeat the procedure after switching the samples each time for a minimum of a total four irradiations, or more if high accuracy is desired. Sum up the individual counts and take their ratio.

Note that the effect of *dead time* (p.220) can be ignored if both samples have similar compositions, and modern multichannel analyzers compensate for this effect internally. Working at relatively low counting rates minimizes the dead-time effect, which can be significant if samples of widely differing composition are compared.

In this procedure, the advantages of a dual transfer system and parallel simultaneous counting are even more obvious. For example, in a single tube transfer system, one has to normalize for variations in neutron flux, irradiation time, and decay time, which involves calculations taking such decay factors into account as $1 - e^{-\lambda t}$, and $e^{-\lambda t}$. When simultaneous accumulation of data from two samples can not be realized, problems of correction for the different relative starting times ($t = 0$) arise.

Note that phosphorus-31 interferes here because of reaction:

$$^{31}P(n,\alpha)^{28}Al$$

In accurate work the concentration of phosphorus must, therefore, be known and its contribution to the total intensity must be estimated and corrected for.

Gravimetric Factors

Ag	$= AgCl \times 0.75263$		Mn	$= Mn_2P_2O_7 \times 0.38713$
Al	$= Al_2O_3 \times 0.52923$		MnO	$= Mn_2P_2O_7 \times 0.49988$
Al	$= Al(C_9H_6NO)_3 \times 0.05872$		Na	$= Na_2O \times 0.74186$
Al_2O_3	$= Al(C_9H_6NO)_3 \times 0.11096$		Na	$= NaZn(UO_2)_3(C_2H_3O_2)_9 \cdot$
B	$= B_2O_3 \times 0.31057$			$\cdot 6H_2O \times 0.014948$
Ba	$= BaO \times 0.89566$		NaCl	$= NaZn(UO_2)_3(C_2H_3O_2)_9 \cdot$
Ba	$= BaSO_4 \times 0.58845$			$\cdot 6H_2O \times 0.038001$
BaO	$= BaSO_4 \times 0.65700$		Na_2O	$= NaZn(UO_2)_3(C_2H_3O_2)_9 \cdot$
C	$= CO_2 \times 0.27291$			$\cdot 6H_2O \times 0.020149$
Ca	$= CaO \times 0.71469$		Ni	$= NiO \times 0.78584$
Ca	$= CaC_2O_4 \cdot H_2O \times 0.27430$		Ni	$= Ni[(CH_3)_2(CNO)_2H]_2 \times$
CaO	$= CaC_2O_4 \cdot H_2O \times 0.38379$			$\times 0.20319$
Cl	$= AgCl \times 0.24737$		NiO	$= Ni[(CH_3)_2(CNO)_2H]_2 \times$
Cu	$= CuO \times 0.79884$			$\times 0.25857$
CuO	$= Cu \times 1.2518$		O	$= F \times 0.42107$
Cu_2O	$= Cu \times 1.1259$		P	$= P_2O_5 \times 0.43642$
F	$= CaF_2 \times 0.48664$		P	$= Mg_2P_2O_7 \times 0.27831$
F	$= CaSO_4 \times 0.27911$		P	$= (NH_4)_3PO_4 \cdot 12MoO_3 \times$
F	$= PbClF \times 0.07263$			$\times 0.016507$
Fe	$= Fe_2O_3 \times 0.69944$		P_2O_5	$= Mg_2P_2O_7 \times 0.63776$
FeO	$= Fe_2O_3 \times 0.89981$		P_2O_5	$= (NH_4)_3PO_4 \cdot 12MoO_3 \times$
Fe_2O_3	$= FeO \times 1.1113$			$\times 0.037823$
H	$= O \times 0.12600$		Pb	$= PbO \times 0.92832$
H	$= H_2O \times 0.11190$		Pb	$= PbSO_4 \times 0.68324$
K	$= K_2O \times 0.83015$		Pb	$= PbCrO_4 \times 0.64108$
K	$= KClO_4 \times 0.28222$		PbO	$= PbSO_4 \times 0.73600$
K	$= KB(C_6H_5)_4 \times 0.10912$		PbO	$= PbCrO_4 \times 0.69061$
KCl	$= K \times 1.9068$		S	$= SO_3 \times 0.40049$
KCl	$= KClO_4 \times 0.53810$		S	$= BaSO_4 \times 0.13737$
K_2O	$= K \times 1.2046$		SO_3	$= BaSO_4 \times 0.34300$
K_2O	$= KClO_4 \times 0.33993$		Si	$= SiO_2 \times 0.46744$
Mg	$= MgO \times 0.60311$		Zn	$= ZnO \times 0.80339$
Mg	$= Mg_2P_2O_7 \times 0.21847$		Zr	$= ZrO_2 \times 0.74030$
MgO	$= Mg_2P_2O_7 \times 0.36224$		ZrO_2	$= ZrP_2O_7 \times 0.46468$
Mn	$= MnO \times 0.77446$			

Analytical Quality Reagents for Stock Solutions in Atomic-Absorption and Emission-Flame Analysis

Buy commercially available 500 or 1,000 p.p.m. standard chloride solutions of the following elements or make these from reagents listed below.

Aluminum chloride, $AlCl_3$	500 g
Aluminum wire, Al, high purity	500 g
Ammonium chloride, NH_4Cl	1,000 g
Barium chloride, $BaCl_2$	500 g
Boric acid anhydride, B_2O_3	1,000 g
Calcium carbonate, anhydrous, $CaCO_3$	1,000 g
Chromium, metal, Cr, high purity	100 g
Copper, Cu, high purity	500 g
Iron, metal, Fe, high purity	500 g
Lanthanum oxide, La_2O_3	1,000 g
Lead, Pb, high purity	500 g
Magnesium metal, Mg, high purity	100 g
Manganese metal, Mn, high purity	100 g
Nickel, Ni, high purity	500 g
Potassium chloride, KCl	1,000 g
Sodium carbonate, anhydrous, Na_2CO_3	1,000 g
Sodium chloride, NaCl	1,000 g
Sodium metasilicate, Na_2SiO_3	1,000 ml
Strontium chloride, $SrCl_2$	500 g
Titanium oxide, TiO_2	500 g
Zinc metal, Zn, high purity	500 g
Hydrochloric acid, conc., HCl, 1.18 s.g.	5,000 ml
Hydrofluoric acid, HF, 40%	1,000 ml
Perchloric acid, conc., $HClO_4$, 70%	1,000 ml
Nitric acid, conc., HNO_3, 1.42 s.g.	1,000 ml
Sulfuric acid, conc., H_2SO_4, 1.84 s.g.	5,000 ml
Acetic acid, CH_3COOH, glacial	1,000 g
Ethyl alcohol, C_2H_5OH, 95%	5,000 ml
Hydrogen peroxide, H_2O_2, 30%	1,000 ml

Fuel:

Acetylene, C_2H_2	2 cylinders
Nitrous oxide, N_2O, as oxidizer	2 cylinders

Bibliography

Selected Comprehensive Texts in Classical Analytical Chemistry

BELCHER, R. and NUTTEN, A. J., 1960. *Quantitative Inorganic Analysis*, 2nd ed. Butterworths, London, 390 pp.

BENNETT, H. and HAWLEY, W. G., 1965. *Methods of Silicate Analysis*, 2nd ed. Acad. Press, London, 350 pp.

COTTON, F. A. and WILKINSON, G., 1962. *Advanced Inorganic Chemistry*. Interscience (Wiley), New York, N.Y., 959 pp.

ERDEY, L., 1963. *Gravimetric Analysis*. Pergamon, Oxford, 320 pp.

FURMAN, N. H., 1962. *Standard Methods of Chemical Analysis*, 6th ed. Van Nostrand, Princeton, N.J., 1: 1500 pp.; 2: 2400 pp.

GROVES, A. W., 1951. *Silicate Analysis*, 2nd ed. Allen and Unwin, London, 336 pp.

HILLEBRAND, W. F., BRIGHT, H. A., HOFFMAN, J. I. and LUNDELL, G. E. F., 1953. *Applied Inorganic Analysis*, 2nd ed. Wiley, New York, N.Y., Chapman and Hall, London, 1034 pp.

KODAMA, K., 1963. *Methods of Quantitative Inorganic Analysis, an Encyclopedia of Gravimetric, Titrimetric, and Colorimetric Methods*. Interscience (Wiley), New York, N.Y., 544 pp.

KOLTHOFF, I. M. and SANDELL, E. B., 1955. *Textbook of Quantitative Inorganic Analysis*, 3rd ed. Macmillan, New York, N.Y., 759 pp.

LAITINEN, H. A., 1960. *Chemical Analysis*. McGraw-Hill, New York, N.Y., 611 pp.

LUNDELL, G. E. F. and HOFFMAN, J. I., 1938. *Outlines of Methods of Chemical Analysis*. Wiley, New York, N.Y., 250 pp.

MEITES, L., 1963. *Handbook of Analytical Chemistry*. McGraw-Hill, New York, N.Y., 1788 pp.

MEURICE, A. et MEURICE, C., 1954. *Analyse des Matières Minérales*, 4me ed. Dunod, Paris, 982 pp.

VOINOVITCH, I. A., DEBRAS-GUÉDON, J. et LOUVRIER, J., 1962. *L'analyse des Silicates*. Hermann, Paris, 512 pp.

WILSON, C. L. and WILSON, D. W., 1959, 1960, 1962. *Comprehensive Analytical Chemistry*. Elsevier, Amsterdam, 1a (1959): 578 pp.; 1b (1960): 878 pp.; 1c (1962): 728 pp.

Special Texts

ADLER, I., 1966. *X-ray Emission Spectrography in Geology*. Elsevier, Amsterdam, 258 pp.

AHRENS, L. H. and TAYLOR, S. R., 1961. *Spectrochemical Analysis. A Treatise on the d-c Arc Analysis of Geological and Related Materials*, 2nd ed. Addison-Wesley, Reading, Mass., 454 pp.

ALDER, H. L. and ROESSLER, E. B., 1962. *Introduction to Probability and Statistics*. Freeman, San Fransisco, Ca., 313 pp.

Annual Reviews, Analytical Chemistry. Yearly April editions.

Atomic Absorption Newsletter, 1962. Perkin Elmer Corp., Norwalk, Conn.

BAIRD, A. K. and HENKE, B. L., 1965. Oxygen determinations in silicates and total major elemental analysis of rocks by soft X-ray spectrometry. *Anal. Chem.*, 37: 727–729.

BAUMAN, R. P., 1962. *Absorption Spectroscopy*. Wiley, New York, N.Y., 611 pp.

BIRKS, L. S., 1959. *X-ray Spectrochemical Analysis*. Interscience, New York, N.Y., 144 pp.

BOLTZ, D. F., 1958. *Colorimetric Determination of Nonmetals*. Interscience, New York, N.Y., 384 pp.

BURRIEL-MARTÍ, F. and RAMÍRES-MUÑOZ, J., 1957. *Flame Photometry*. Elsevier, Amsterdam, 531 pp.

DEAN, J. A., 1960. *Flame Photometry*. McGraw-Hill, New York, N.Y., 354 pp.

DEARNALEY, G. and NORTHROP, D. C., 1964. *Semiconductor Counters for Nuclear Radiations*. Spon, London, 331 pp.

DODD, R. E., 1962. *Chemical Spectroscopy*. Elsevier, Amsterdam, 340 pp.

DUVAL, C., 1963. *Inorganic Thermogravimetric Analysis*, 2nd ed. Elsevier, Amsterdam, 722 pp.

ELWELL, W. T. and GIDLEY, J. A. F., 1962. *Atomic Absorption Spectrophotometry*. MacMillan, New York, N.Y., 102 pp.

EWING, G. W., 1960. *Instrumental Methods of Chemical Analysis*. McGraw-Hill, New York, N.Y., 454 pp.

FAIRBAIRN, H. W., and others, 1951. A cooperative investigation of precision and accuracy in chemical, spectrochemical and modal analysis of silicate rocks. *U.S. Geol. Surv. Bull.*, 980: 1–71.

FLASCHKA, H. A., 1959. *EDTA Titrations*. Pergamon, London, 138 pp.

FLEISCHER, M., 1965. Summary of new data on rock samples G-1 and W-1, 1962–1965. *Geochim. Cosmochim. Acta*, 29: 1263–1283.

GOVINDARAJU, K., 1968. Ion exchange dissolution method for silicate analysis. *Anal. Chem.*, 40: 24–26.

HAMILTON, L. F. and SIMPSON, S. G. 1960. *Calculations of Analytical Chemistry*, 6th ed. McGraw-Hill, New York, N.Y., 334 pp.

HEATH, R. L., 1964. *Scintillation Spectrometry*, 2nd. ed. United States Gov. Pr. Off., Washington, D.C.

HELFFERICH, F., 1962. *Ion Exchange*. McGraw-Hill, New York, N.Y., 640 pp.

HERRMANN, R. and ALKEMADE, C. T. J., 1963. *Flame Photometry*. Interscience (Wiley), New York, N.Y., 512 pp.

INGAMELLS, C. O., 1966. Absorptiometric methods in rapid silicate analysis. *Anal. Chem.*, 38: 1228–1234.

KOLTHOFF, I. M., BELCHER, R., STENGER, V. A. and MATSUYAMA, G., 1947. *Volumetric Analysis*, 2, 3. Interscience, New York, N.Y., 2: 388 pp.; 1957, 3: 724 pp.

KRUMBEIN, W. C. and GRAYBILL, F. A., 1965. *An Introduction to Statistical Models in Geology*. McGraw-Hill, New York, N.Y., 512 pp.

LANGMYIR, F. J. and GRAFF, P. R., 1965. A contribution to the analytical chemistry of silicate rocks: a scheme of analysis for eleven main constituents based on decomposition by hydrofluoric acid. *Norg. Geol. Undersøk.*, 230: 1-128.

LIEBHAFSKY, H. A., PFEIFFER, H. G., WINSLOW, E. H. and ZEMANY, P. D., 1960. *X-ray Absorption and Emission in Analytical Chemistry*. Wiley, New York, N.Y., 357 pp.

MAYNES, A. D., 1965. A procedure for silicate rock analysis based on ion exchange and complex ion formation. *Anal. Chim. Acta*, 32: 211–220.

MAYNES, A. D., 1966. A semi-micro procedure for mineral analysis. *Chem. Geol.*, 1 (1): 61–75.

MILLER, R. L. and KAHN, J. S., 1962. *Statistical Analysis in the Geological Sciences*. Wiley, New York, N.Y., 483 pp.

MOENKE, H., 1962. *Spektralanalyse von Mineralien und Gesteinen*. Geest und Portig, Leipzig, 222 pp.

MORRISON, G. H. and FREISER, H., 1957. *Solvent Extraction in Analytical Chemistry*. Wiley, New York; Chapman and Hall, London, 269 pp.

MUNCE, J. F., 1962. *Laboratory Planning*. Butterworth, London, 360 pp.

NALIMOV, V. V., 1963. *The Application of Mathematical Statistics to Chemical Analysis*. Pergamon, London, 294 pp.

OVERMAN, R. T. and CLARK, H. M., 1960. *Radioisotope Techniques*. McGraw-Hill, New York, N.Y., 476 pp.

PECK, L. C., 1964. Systematic analysis of silicates. *U.S. Geol. Surv. Bull.*, 1170: 1–89.

POLUÉKTOV, N. S., 1961. *Techniques in Flame Photometric Analysis*. Consultants Bur., New York, N.Y., 235 pp.

RINGBOM, A., 1963. *Complexation in Analytical Chemistry*. Interscience (Wiley), New York, N.Y., 395 pp.

ROSE, H. J., ADLER, I. and FLANAGAN, F. J., 1963. X-ray fluorescence analysis of the light elements in rocks and minerals. *Appl. Spectry.*, 17 (4): 81–85.

SCHWARZENBACH, G., 1960. *Die Komplexometrische Titration*. Enke, Stuttgart, 119 pp.

SHAPIRO, L. and BRANNOCK, W. W., 1962. Rapid analysis of silicate, carbonate and phosphate rocks. *U.S. Geol. Surv. Bull.*, 1144-A: 1–56.

SKOOG, D. A. and WEST, D. M., 1963. *Fundamentals of Analytical Chemistry*. Holt, Rinehart and Winston, New York, N.Y., 786 pp.

SMALES, A. A. and WAGER, L. R., 1960. *Methods in Geochemistry*. Interscience (Wiley), New York, N.Y., 471 pp.

SNELL, F. D., SNELL, C. T. and SNELL, C. A., 1959. *Colorimetric Methods of Analysis*. Van Nostrand, Princeton, N.J., 804 pp.

STARY, J., 1964. *The Solvent Extraction of Metal Chelates*. Pergamon, Oxford, 240 pp.

STEVENS, R. E. and NILES, W. W., 1960. Second report on a cooperative investigation of the composition of two silicate rocks. *U.S. Geol. Surv. Bull.*, 1113: 1–126.

SUHR, N. H. and INGAMELLS, C. O., 1966. Solution technique for analysis of silicates. *Anal. Chem.*, 38: 730–734.

TAGGART, A. F., 1945. *Handbook of Mineral Dressing*, Wiley, New York, N.Y., 1945 pp.

VOLBORTH, A., 1963. Total instrumental analysis of rocks. *Nevada Bur. Mines, Rept.* 6A: 72 pp.

VOLBORTH, A., 1965. Dual grinding and X-ray analysis of all major oxides in rocks to obtain true composition. *Appl. Spectr.*, 19: 1–7.

VOLBORTH, A., FABBI, B. P. and VINCENT, H. A., 1968. Total nondestructive analysis of CAAS syenite. *Advan. X-ray Anal.*, 11: 158–163.

WALSH, J. W. T., 1958. *Photometry*, 3rd ed. Constable, London, 544 pp.

WARD, F. N., LAKIN, H. W., CANNEY, F. C., 1963. Analytical methods used in geochemical exploration by the U.S. Geological Survey. *U.S. Geol. Surv. Bull.*, 1152: 100.

WELCHER, F. J., 1958. *The Analytical Uses of Ethylenediamine Tetraacetic Acid.* Van Nostrand, Princeton, N.J., 366 pp.

WELDAY, E. E., BAIRD, A. K.. MCINTYRE, D. B. and MADLEM, K. W., 1964. Silicate sample preparation for light-element analyses by X-ray spectrography. *Am. Mineralogist*, 49: 889–903.

Index